D1384316

CONCRETE TECHNOLOGY AND DESIGN

Volume 2

New Reinforced Concretes

CONCRETE TECHNOLOGY AND DESIGN

A series highlighting the importance of the inherent properties of concrete materials and their influence on design. Each volume covers a particular aspect of concrete technology or of design, and contains contributions from internationally acknowledged experts. The series will prove very useful to practising engineers and architects, researchers, and advanced students in civil and structural engineering and materials technology.

Series Editor: R.N. Swamy, PhD, CEng, FICE, FIStructE, FASCE.

Volume 1: New Concrete Materials

CONCRETE TECHNOLOGY AND DESIGN

Volume 2

New Reinforced Concretes

Editor

R.N. Swamy, BEng, MEng, PhD, CEng, FICE, FIStructE, FASCE, FIE (Ind)
Reader in Civil and Structural Engineering
University of Sheffield

Surrey University Press

Published by Surrey University Press
A member of the Blackie Group
Bishopbriggs, Glasgow G64 2NZ and
Furnival House, 14–18 High Holborn,
London WCIV 6BX

First published 1984

British Library Cataloguing in Publication Data

New reinforced concretes.—(Concrete
technology and design; v. 2)
1. Reinforced concrete
I. Swamy, R.N.
624.1'8341 TA444

ISBN 0–903384–47–7

Phototypeset by Thomson Press (India) Ltd., New Delhi
Printed in Great Britain by Bell & Bain (Glasgow) Ltd.

Contents

Contributors

S.P. Shah	Professor of Civil Engineering, Northwestern University, Illinois
P. Balaguru	Associate Professor of Civil Engineering, Rutgers University, New Jersey
J.G. Keer	Lecturer, Department of Civil Engineering, University of Surrey, Guildford
M.A. Aziz	Associate Professor, Department of Civil Engineering, National University of Singapore
P. Paramasivam	Associate Professor, Department of Civil Engineering, National University of Singapore
S.L. Lee	Professor of Civil Engineering, National University of Singapore
B.V. Subrahmanyam	Chief Design Engineer, Associated Constructions, Dubai, and formerly Head, Concrete Composites Laboratory, Structural Engineering Research Centre, Taramani, Madras

Preface

Science and technology do not stand still. They make rapid advances and unforeseen changes in short periods of time, and the construction industry can no more remain immune to or isolated from such changes than society or humanity at large. In the past the role of materials in the performance of structures has either been taken for granted or inadequately appreciated by engineers, architects and builders. Engineers have often no clear idea of the behaviour of materials whereas physicists and material scientists do not understand how the effects of loads and stresses in a structural context modify the performance characteristics of materials. The only way to reduce this unhappy imbalance is for engineers to take a keener interest in the materials they use, and for the material technologists to appreciate the realities of material properties in construction.

This is the second volume in the Series 'Concrete Technology and Design' aimed to bridge the gap between materials science and design. As in volume one, the book is devoted to a single theme and the chapter subjects are selected in such a way that they contribute to the central theme and at the same time interact closely with each other.

This particular volume aims to show that conventional reinforced concrete, as we understand it and as has been used for over a century, is not the only form that reinforced concrete can take. Ferrocement and fibre reinforced concrete, which form the basis of this book are neither new to engineers nor are they in reality new forms of construction, although engineers seem to have forgotten the pioneering role played by these two materials in the early history of concrete construction.

The use of ferrocement dates back to the 1850s but it is only lately that engineers have recognized the potential of ferrocement construction in structural applications not merely to developing countries but also to developed technologies. The use of fibre reinforcement also goes back for nearly a century, and the most successful commercial application followed the invention of asbestos cement by Hatschek just before the beginning of this century. Since then fibre concrete in bulk form and as thin sheet products have been developed. Current evidence shows that fibre reinforcement can be exploited further in conjunction with both reinforced and prestressed concrete.

The use of natural fibres and bamboo reinforcement has special relevance to developing and Third World countries. Their natural abundance, inexhaustible source and relative cheapness provide an alternative reinforcing medium not only for cement matrices but also for soil cement and ferrocement type of construction. The first recorded use of bamboo reinforcement was in China in 1919, and several such constructions are reported to have been made both in the USA and Japan during the Second World War. Their areas of application are limitless and not merely confined to housing and disaster or refugee relief structures.

The authors of the four chapters are well known for their research contributions, and have made a special effort to relate the material properties to design and construction. I hope that their excellent effort will enable engineers, architects, builders, and particularly the new generation of young engineers to think beyond conventional concrete forms, and in terms of available materials and the needs of society.

R N Swamy

1 Ferrocement

S.P. SHAH and P.N. BALAGURU

Abstract

Ferrocement is a type of thin reinforced concrete construction where usually a hydraulic cement is reinforced with layers of continuous and relatively small-diameter wire meshes. In the past 20 years there has been increasing field application and laboratory research with this type of construction. This chapter is intended to provide the basic state-of-the-art knowledge for both designers and researchers.

The chapter is divided into six sections dealing with: the definition and introduction, history of development, mechanical behaviour, recommendations for material selection and design, applications and conclusions. The related list of references is presented at the end of the chapter. Properties of ferrocement in compression, tension and flexure are presented in the mechanical behaviour section. Design recommendations provide guidelines regarding allowable stresses and serviceability criteria. A design example is worked out to clarify the analysis procedure for axial forces and bending. Applications of ferrocement for construction of boats, silos, tanks and roof shells are discussed.

1.1 Introduction

Ferrocement can be considered as a modified form of reinforced concrete. The major differences between a conventional reinforced concrete structural element and a ferrocement member can be enumerated as follows.

(i) Ferrocement structural elements are normally thin, thickness rarely exceeding 25 mm. Conventional reinforced concrete members, on the other hand, consist of relatively thick sections with thicknesses often exceeding 100 mm.

(ii) Matrix in ferrocement mainly consists of portland cement mortar instead of regular concrete which contains coarse aggregate.

(iii) The reinforcement provided in ferrocement consists of large amounts of smaller-diameter wires or wire meshes instead of discretely-placed reinforcing bars used in reinforced concrete. Moreover, ferrocement normally contains a greater percentage of reinforcement, distributed throughout the cross-section.

(iv) In terms of structural behaviour, ferrocement exhibits very high tensile strength to weight ratio and superior cracking performance.

(v) In terms of construction, formwork is very rarely needed (or used) for the fabrication. This aspect permits economical construction of certain structures such as geodesic domes, wind tunnels, circular storage structures and swimming pools.

In short, *ferrocement can be considered a type of thin reinforced concrete construction where instead of discretely-placed reinforcing bars, large amounts of smaller diameter wire meshes are used uniformly throughout the cross-section, and instead of concrete, portland cement mortar is used.*

Metallic mesh is the most common type of reinforcement. Meshes made of alkali-resistant glass fibres and woven fabric made of vegetable fibres such as jute burlap and bamboo have also been tried as reinforcement[1,2]. Sometimes, regular reinforcing bars in a skeletal form are added to the thin wire meshes in

Table 1.1 Ferrocement: current ranges of composition and properties

Wire-mesh performance	
Wire diameter:	$0.020 \leqslant \phi \leqslant 0.062$ in ($0.5 \leqslant \phi \leqslant 1.5$ mm)
Type of mesh:	Chicken wire or square woven- or welded-wire galvanized mesh; expanded metal
Size of mesh openings:	$\frac{1}{4} \leqslant m \leqslant 1$ in ($6 \leqslant m \leqslant 25$ mm)
Number of mesh layers:	Up to 12 layers per in. of thickness (up to 5 layers per cm of thickness)
Fraction volume of reinforcement:	Up to 8% in both directions corresponding to up to 40 pounds of steel per cubic foot of concrete (630 kg/m^3)
Specific surface of reinforcement:	Up to 10 in^2/in^3 in both directions (up to 4 cm^2/cm^3 in both directions)
Intermediate skeletal reinforcement (if used)	
Type:	Wires; wire fabric, rods; strands
Diameter:	$\frac{1}{8} \leqslant d \leqslant \frac{3}{8}$ in ($3 \leqslant d \leqslant 10$ mm)
Grid size:	$2 \leqslant G \leqslant 4$ in ($5 \leqslant G \leqslant 10$ cm)
Typical mortar composition	
Portland cement:	Any type depending on application
Sand-to-cement ratio:	$1 \leqslant S/C \leqslant 2.5$ by weight
Water-to-cement ratio:	$0.4 \leqslant W/C \leqslant 0.6$ by weight
Recommendations:	Fine sand all passing U.S. sieve No. 8 and having 5% by weight passing No. 100, with a continuous grading curve in between
Composite properties	
Thickness:	$\frac{1}{4} \leqslant t \leqslant 2$ in ($6 \leqslant t \leqslant 50$ mm)
Steel cover:	$\frac{1}{16} \leqslant c \leqslant \frac{3}{16}$ in ($1.5 \leqslant c \leqslant 5$ mm)
Ultimate tensile strength:	Up to 5000 psi (34.5 MPa)
Allowable tensile stress:	Up to 1500 psi (10.3 MPa)
Modulus of rupture:	Up to 8000 psi (55.1 MPa)
Compressive strength:	4000 to 10 000 psi (27.6 to 68.9 MPa)

order to achieve a stiff reinforcing cage[3]. Naturally, the contribution of these extra reinforcements should be considered in the analysis and design. On rare occasions, ferrocement is also used in the prestressed form.[3,4] The commonly used composition and properties of ferrocement made with steel wire mesh reinforcement are summarized in Table 1.1. This chapter primarily deals with ferrocement reinforced with steel wire meshes.

1.2 History of development

The ferrocement as defined in the preceding section is generally credited to Nervi and dates from the work that he did in the years 1942–43.[5–7] It is interesting to note that one of the first applications of reinforced concrete construction was the ferrocement rowing boat built by Lambot in France in 1849. Lambot took out French and Belgian patents on what he termed 'Ferciment' in 1856. One of his boats was still afloat 100 years later in 1949 and is currently on display in a museum in Brignoles, France, reportedly in good condition[8]. In 1887, a similar boat was constructed in Holland. This vessel, now nearly 100 years old, is still afloat on the Pelican Pond at Amsterdam Zoo[9].

There was very little application of true ferrocement construction between 1888 and 1942 when Pier Luigi Nervi began a series of experiments on ferrocement. He observed that reinforcing concrete with layers of wire mesh produced a material possessing the mechanical characteristics of an approximately homogeneous material capable of resisting high impact. Thin slabs of concrete reinforced in this manner proved to be flexible, elastic and exceptionally strong. After the Second World War, Nervi demonstrated the utility of ferrocement as a boat-building material by building the 165-ton motor sailer *Irene* with a 3.6-cm-thick ferrocement hull. He reported that the weight of the vessel was approximately 5% less than an equivalent wooden ship and its cost was 40% as compared to a similar wooden hull.

Nervi also applied the ferrocement concept to civil engineering structures. He built a small warehouse whose walls and roof were 3-cm-thick corrugated ferrocement. He used the concept of corrugation for the roofs of several major structures including a roof system spanning 98 m for the Turin Exhibition Hall. The ferrocement corrugated units were less than 4 cm thick, were precast and joined by reinforced concrete ribs cast *in situ* in the troughs and crests of the corrugations[7].

Despite this evidence that ferrocement was an adequate and economic building material, it gained wide acceptance only in the early 1960s. In 1965, an American-owned ferrocement yacht built in New Zealand, the 17 m *Awanee*, circumnavigated the world without any serious mishap, although it encountered gales of 70 knots, collided with an iceberg, and was rammed by a steel-hull yacht. Since then, there has been increasing activity with ferrocement construction throughout the world including Canada, the U.S.A., Australia,

New Zealand, the U.K., Mexico, the Soviet Union, Poland, the Republic of China, Thailand, India, Indonesia, and many other developing countries. The applications of ferrocement construction include, in addition to boat hulls, floating marine structures, roofs, silos, pipes, water tanks and low-cost housing.

In 1972, the U.S. National Academy of Sciences, through its Board on Science and Technology for International Development, established an *ad hoc* panel on the Utilization of Ferrocement in Developing Countries. The report of the panel is a good summary of the history and applications of ferrocement construction[10].

In 1974, the American Concrete Institute formed Committee 549 on Ferrocement. The mission of the committee was to develop a body of knowledge on the engineering properties, construction practices and practical applications of ferrocement and similar materials. From time to time the committee organizes and co-sponsors international conferences in order to collect and disseminate information for design engineers around the world. The committee's recent publications include a symposium proceedings[11] and a state-of-the-art[12] report.

In 1976, the International Ferrocement Information Centre (IFIC) was founded at the Asian Institute of Technology (P.O. Box 2754, Bangkok, Thailand). The centre is financed by the United States Agency for International Development, the government of New Zealand and the International Development Research Center of Canada. The centre serves as a clearing house for information on ferrocement and publishes the *Journal of Ferrocement*. Sometimes a particular issue of the journal is dedicated to a certain application. For instance, issue No. 3 of Volume 10 (1980) was dedicated to Marine Applications. The centre also conducts workshops and sponsors international conferences. The recent publications of the centre include a book[13], a conference proceedings[14], and booklets in the 'do-it-yourself' series[15].

In 1979, RILEM (International Union of Testing and Research Laboratories of Materials and Structures) established a committee (48-FC) to evaluate testing methods for ferrocement. The committee sponsored a recent international symposium[16,17] on some aspects of the behaviour, design and applications of ferrocement.

1.3 Mechanical behaviour

During the late sixties and seventies a considerable amount of research has been conducted on the mechanical properties of ferrocement[18–45]. The information presented in this section is based on these extensive studies. The discussion primarily deals with ferrocement reinforced with welded or woven square meshes. The current state-of-the-art knowledge on the use of hexagonal chicken wire mesh and expanded metal lath is not sufficient to formulate a clearcut design recommendation. However, the design engineer can use the current literature prudently. References 12, 19, 20, 30 and 46 provide some relevant information.

1.3.1 *Behaviour under compression*

For most structural applications where thin ferrocement elements are used, the properties of the matrix, namely the portland cement mortar, control the properties of the ferrocement in compression. In other words, thin ferrocement plate elements can be treated as plain mortar plates for most practical applications. The addition of wire meshes does increase the stiffness; however, the increase is not significant enough to be considered in design. In special circumstances where the reinforcement is very heavy, the contribution of reinforcement could be considered in the design, as explained in the design example. Further discussion is not included here because very rarely is the reinforcement contribution considered for compression loading.

To the best of the authors' knowledge, studies are not available for the behaviour of ferrocement under fatigue compression. One could expect that ferrocement will have a better fatigue resistance than the plain mortar for which studies are available[23]. Hence, as a conservative design approach, one can neglect the contribution of the wire meshes and design the structural element using the fatigue resistance of the mortar alone.

1.3.2 *Behaviour under tension*

The distinct behaviour of ferrocement can be more clearly understood by studying the uniaxial stress–strain curve of ferrocement specimens. Several investigators have reported that when a ferrocement specimen is subjected to increasing tensile stresses, three stages of behaviour are observed (Fig. 1.1).

In the first stage the composite behaves like a linear elastic solid. This elastic behaviour stage is terminated by the occurrence of the first crack in the matrix.

The second stage can be termed as the multiple cracking stage. Theoretically, this stage starts at the occurrence of the first crack in the matrix and continues up to the point where wire meshes start to yield. In this range of loading, the number of cracks keeps increasing with an increase in tensile stress or strain. However, the crack widths increase very little. The increases in strains, which occur under larger loads, are distributed through a greater number of cracks instead of widening the existing cracks. The number of cracks and their widths depend on a number of parameters such as type and amount of reinforcement, as explained subsequently. It is important to note that this multiple cracking stage represents the behaviour of most structural elements under service (working) loads because the service loads are always large enough to induce cracking but seldom large enough to produce yielding of steel.

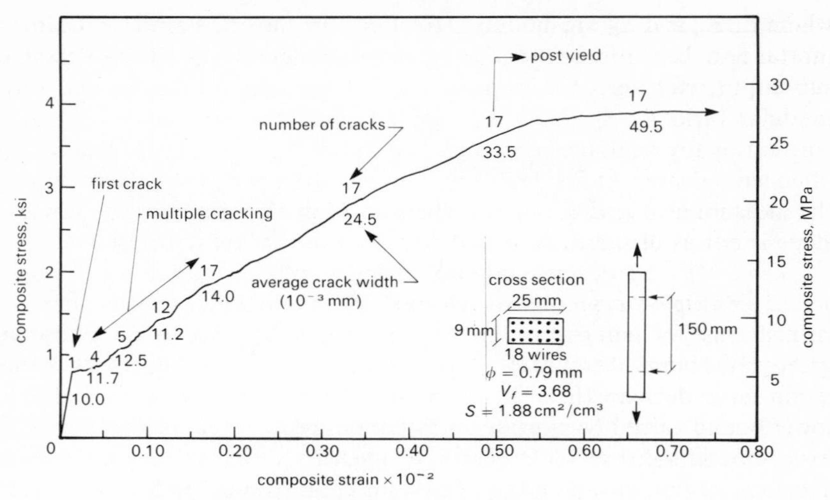

Figure 1.1 Stress–strain curve of ferrocement in tension

In the third stage, which can be called a crack-widening stage, no (or few) additional cracks are formed. This stage begins when the reinforcement starts to yield and continues up to the failure. The number of cracks remains essentially constant and the crack widths keep increasing. The behaviour of the specimen is primarily controlled by that of the reinforcement alone.

The three stages described above can be clearly seen in Fig. 1.1 where the stress–strain curve of a specimen reinforced with a large number of very small-diameter rods is shown[21,22]. The elongation (and equivalently the strain) were measured by a device directly attached to the specimen and equipped with a linear voltage differential transformer (LVDT). Cracks along the observed gauge length were observed with a very sensitive microscope with an accuracy of about 1.27×10^{-3} mm. Figure 1.1 shows that during the multiple cracking stage the number of cracks increased from 1 to 17 while the stress in the longitudinal steel varied from about 138 to 345 MPa; the average crack width, however, increased only slightly from 0.01 mm to 0.014 mm. The steel stress at which the number of cracks stops increasing is termed the stabilization stress. Depending upon the reinforcing parameters, this value may be somewhere along the elastic or non-linear range of behaviour of the steel reinforcement. Each of these three stages is discussed in detail in what follows.

1.3.2.1 *Elastic behaviour.* In the elastic stage, the stress–strain relationship can be defined using a single elastic constant, namely the Young's modulus of the ferrocement composite. The modulus of the ferrocement which will be a function of the moduli of mortar and reinforcement and their relative volume fractions can be expressed as:

$$E_c = (1 - V_{RL})E_m + V_{RL}E_R \simeq E_m + V_{RL}E_R = E_m(1 + V_{RL}n) \qquad (1.1)$$

where E_c, E_m and E_R are moduli of elasticity for the composite (ferrocement), mortar and the reinforcement, V_{RL} is the volume fraction of reinforcement, the subscript L referring to the properties in the longitudinal direction, and n is the modular ratio E_R/E_m. Note that eq. (1.1) is of the same form as that for conventionally reinforced concrete. The end of the elastic linear behaviour is often termed stress at first cracking. This value depends upon the accuracy of the measurement and quite often the transition from the first to the second stage is not as distinctly observed as shown in Fig. 1.1.

1.3.2.2 *Multiple cracking stage.* During the multiple cracking stage, the contribution of mortar to the stiffness of the composite decreases progressively. Hence, the stiffness of the composite, and the slope of the stress–strain curve, decreases from an upper bound value represented by eq. (1.1) to a lower bound value in which the contribution from mortar becomes zero. The lower bound value of the modulus of elasticity can be obtained by simply substituting zero for E_m in eq. (1.1). Hence, if one wants to use a conservative approach in design, the stress–strain relationship of the ferrocement in the multiple cracking stage can be approximately represented by the quation:

$$E_c = V_{RL} E_R \tag{1.2}$$

For a given stress, the strains predicted by using E_c of eq. (1.2) will always be greater than or equal to the actual strains.

The value of E_R may be substantially different for woven mesh from that for welded mesh. The value of the modulus is also found to be different if measured along the transverse direction instead of the longitudinal direction. For welded mesh, the value of E_R is approximately the same as that of the steel (200 GPa). For woven mesh, E_R can be substantially lower than that depending upon the pitch of the weave and other parameters. For the types of square woven meshes generally used in ferrocement E_R can vary between 103 and 200 GPa.

1.3.2.3 *Cracking behaviour.* It has been observed that, everything else being equal, the higher the volume of reinforcement and smaller the diameter of the wires, the larger the extent of multiple cracking stage, the larger the number of cracks developed in the same gauge length, the smaller the final crack spacing and the smaller the crack width. A parameter which is found critical in determining the cracking behaviour is the specific surface of reinforcement, S_R, which is defined as the lateral surface (bonded area) of the reinforcement per unit volume of the composite. S_R is related to the volume fraction of steel as follows:

$$S_R = \frac{4V_R}{\varphi} \tag{1.3}$$

where φ is the diameter of the wire used.

A direct relationship between the average crack spacing $(\Delta l)_{av}$ at crack stabilization and the specific surface of reinforcement seems to exist. An analytical evaluation of this relationship has been proposed[19] and is given by

$$(\Delta l)_{av} = \frac{\theta}{\eta} \frac{1}{S_{RL}} \tag{1.4}$$

where $(\Delta l)_{av}$ = average crack spacing
θ = a factor relating average crack spacing to maximum crack spacing
η = ratio of bond strength to matrix tensile strength
S_{RL} = specific surface of reinforcement in the loading direction.

Since the factor θ/η can be approximated for design purposes, a direct relation between $(\Delta l)_{av}$ and S_{RL} can be obtained. The observed values of average crack spacing at crack stabilization are shown (Fig. 1.2) versus the specific surface or reinforcement in the loading direction S_{RL} for two series of specimens; one series reinforced with smooth longitudinal wires and the second series reinforced with square wire meshes[21,22]. It can be seen that for the same specific surface, smaller spacings are obtained when the reinforcement contains transverse wires (i.e. meshes). The analytical equation (1.4) is also plotted with the values of $(\theta/\eta) = 1$ and 2.7. It can be observed that for $(\theta/\eta) = 1$,

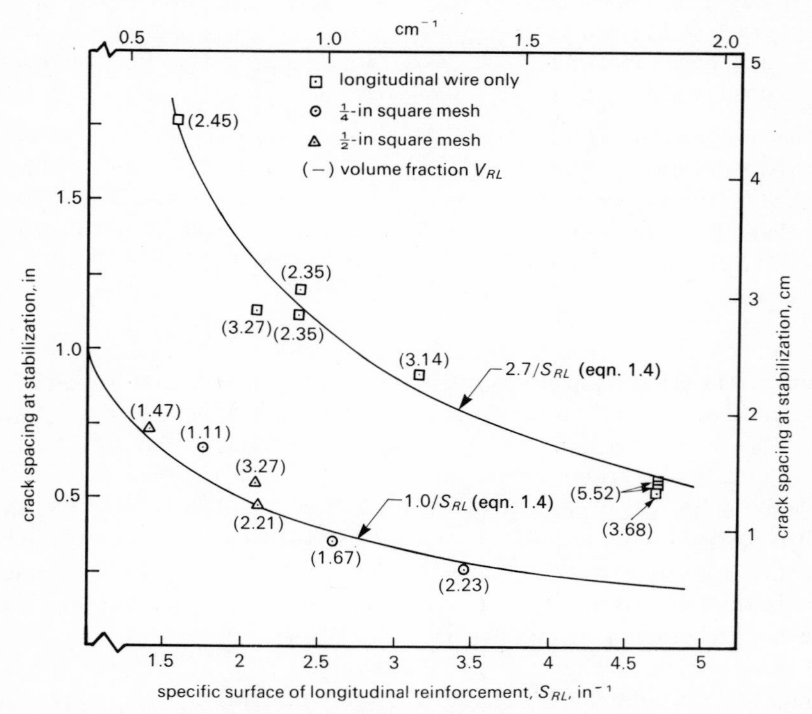

Figure 1.2 Variations in final crack spacing with the specific surface of the reinforcement

eq. (1.4) predicts very well the experimental data for square meshes while for $(\theta/\eta) = 2.7$ it predicts the data when smooth longitudinal wires are used. Thus, eq. (1.4) can be used as a design equation to predict crack spacing in ferrocement at crack stabilization.

Based on the results of various researchers[11,18–22] and the discussion presented above, one can make the following observations. These observations are used for formulating a crack width prediction equation presented subsequently.

(i) In the multiple cracking stage, the increase in crack width is negligible. Hence, the crack width can be assumed to be constant in the crack stabilization stage.
(ii) The average crack spacing $(\Delta l)_{av}$ at crack stabilization stage can be expressed as a function of the specific surface of the reinforcement.
(iii) The steel stress at which crack stabilization stage ends, keeps increasing with specific surface.
(iv) If the specimen is stressed beyond the crack stabilization stage, the crack widths will increase in proportion to the increase in steel stress.
(v) The average crack width is a function of steel strain (stress) and average crack spacing.
(vi) Based on the experimental results available, the randomness of the crack widths can be assumed to follow the normal distribution curve and, hence, maximum crack width can be calculated using the average crack width.

The aforementioned observations are used to arrive at the following approximate empirical design procedure to predict maximum crack width in cracked tensile ferrocement members. The stress in steel f_s should always be less than the yield strength and in any case less than 400 MPa. For any $f_s \leqslant 345\, S_{RL}$

$$W_{max} = \frac{3500}{E_R} \tag{1.5}$$

where f_s is in MPa, S_{RL} in cm^{-1}, W_{max} in mm and E_R in MPa. For $f_s > 345\, S_{RL}$

$$W_{max} = \frac{20}{E_R}[175 + 3.69(f_s - 345\, S_{RL})]. \tag{1.6}$$

Crack width, W_{max}, versus steel stress, f_s, calculated using eqs. (1.5) and (1.6), is plotted in Fig. 1.3 for a series of six identical specimens[21,22]. It can be seen that for stresses below 400–450 MPa (60–65 ksi) (the stress at crack stabilization) there is little increase in crack width. This means that very high values of design stress can be used for ferrocement structures with crack width still being very low. This is due to the relatively high value of the specific surface and the consequent larger extent of the multiple cracking stage. The specific surface for the conventionally reinforced concrete is about one-tenth

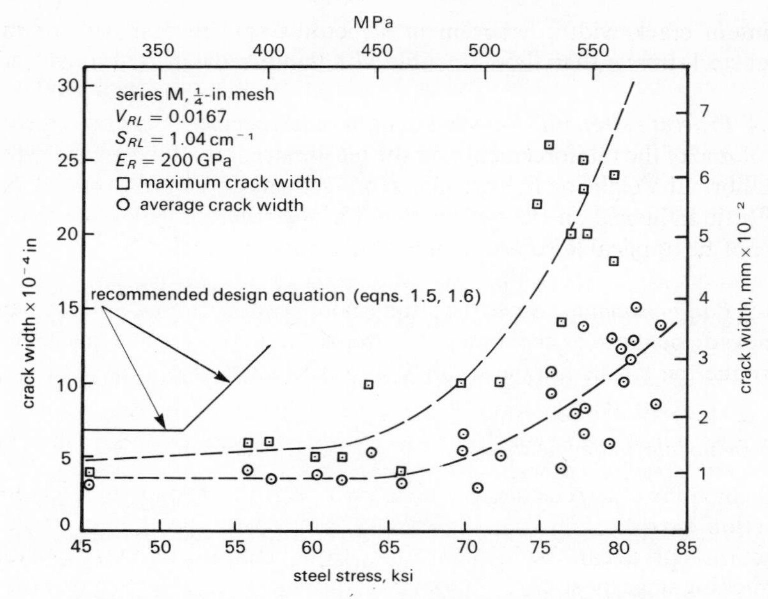

Figure 1.3 Variation of crack width with the steel stress in tension

of that for ferrocement. As a result, much higher design stresses in the reinforcing steel can be used for ferrocement. This can be seen in Fig. 1.4. The data for ferrocement are for cylindrical tanks subjected to internal water pressure[34]. Design of structures such as tanks and silos is based on allowable maximum crack width. It can be seen from Fig. 1.4 that for the same allowable

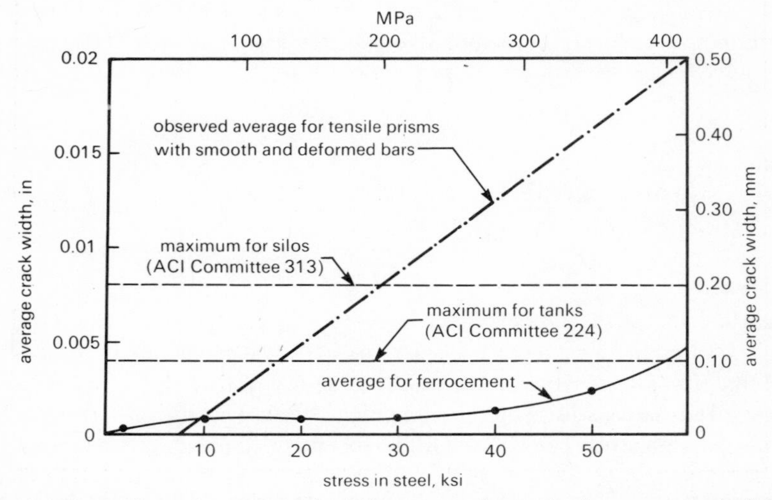

Figure 1.4 Comparison of average crack width versus steel stress for ferrocement and reinforced concrete[34]

maximum crack width, ferrocement structures can be designed for much higher steel stresses than those possible with reinforced concrete construction.

1.3.2.4 *Fracture strength.* Tensile strength of ferrocement depends chiefly on the volume of the reinforcement and the tensile strength of the mesh. Types of sand, normal weight or lightweight, sand–cement and water–cement ratios have little influence on the tensile strength of ferrocement[19]. Note that this may not be applicable to hexagonal or chicken-wire mesh[19].

1.3.2.5 *Fatigue loading.* Since the fatigue failure will occur in reinforcement at a cracked location of the member, fatigue strength of reinforcement will determine the tensile fatigue strength of ferrocement.

1.3.3 *Behaviour under flexure*

The behaviour of ferrocement in flexure can be studied using a typical load–deflection curve[45] (Fig. 1.5). As in the case of tension behaviour, the load–deflection curve can be divided into three regions or stages, namely: precracking stage, post-cracking stage and post-yielding stage. The evaluation of strength measured in terms of resisting moment, stiffness measured in terms of deflection, and cracking behaviour for all the three stages of loading are presented below. The entire discussion on flexure is based on the following assumptions verified by a number of researchers[25–33,35–37].

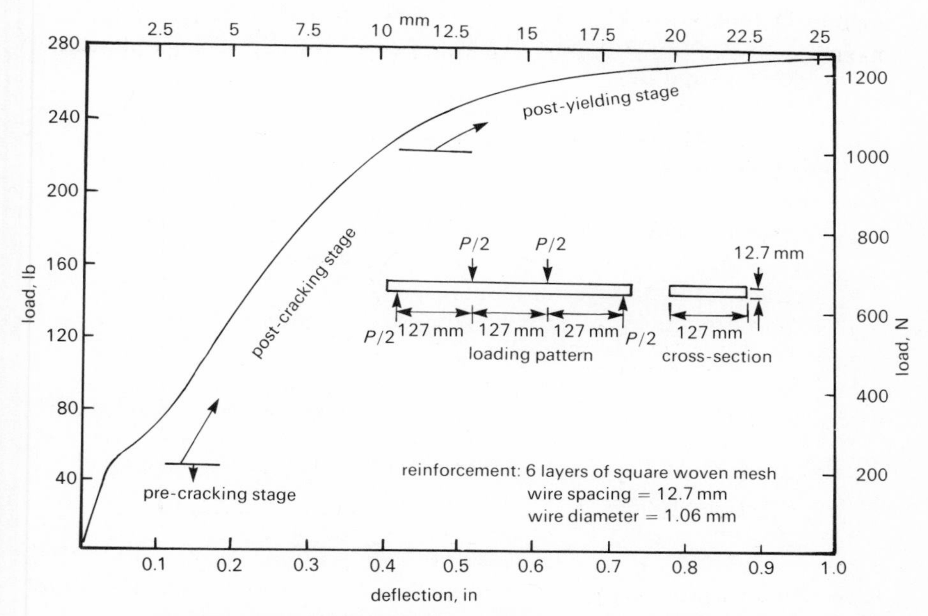

Figure 1.5 A typical load–deflection curve for a ferrocement beam

(i) Plane sections remain plane and perpendicular to the neutral axis, i.e., the strains in mortar and reinforcement are directly proportional to their distance from the neutral axis.
(ii) The behaviour of reinforcement is elastic-perfectly plastic; i.e., for stresses less than yield strength, steel stress is proportional to strain and after yielding stress in steel remains constant at the yield strength f_y.
(iii) Tensile strength of mortar is neglected in flexural strength calculations of cracked beams.
(iv) Maximum usable compression fibre mortar strain is 0.003 mm/mm.
(v) For strength calculations at ultimate load, the parabolic stress–strain distribution of mortar can be approximated to a rectangular distribution using the procedure recommended by the American Concrete Institute Code[47].

1.3.3.1 *Precracking stage.* Ferrocement has the highest stiffness in the precracking stage. In this stage mortar contributes to both compressive and tensile resistance of the composite[45]. The strength and stiffness of the beam can be calculated using the classical bending theory.

The strain, stress and force distributions across the thickness of the beam are presented in Fig. 1.6, in which the following nomenclature is used.

b = width of the beam
h = total thickness of the beam
A_{si} = the total area of the ith layer of reinforcement
d_i = distance of the ith layer of reinforcement from the extreme compression fibre
ε_c = strain in the extreme compression fibre
ε_t = strain in the extreme tension fibre
C_m = total compressive force contribution of mortar
C_s = total compressive force contribution of reinforcement
T_m = total tensile force contribution of mortar
T_s = total tensile force contribution of reinforcement
f_{si} = stress in the ith layer of reinforcement
f_m = mortar stress in the extreme compression or tension fibre
E_m = Young's modulus of mortar
E_R = Young's modulus of reinforcement
jd_i = lever arm of ith layer of reinforcement, measured from neutral axis.

Using the force and moment equilibrium (Fig. 1.6), the resisting moment of the cross-section, M, can be written as:

$$M = C_m(\text{or } T_m) \times \tfrac{2}{3}h + \sum_{i=1}^{m} A_{si} f_{si} jd_i \tag{1.7}$$

where m is the number of layers of reinforcement. Since $C_m = \frac{1}{4}bhf_m$, eq. (1.7) can be rewritten as:

$$M = \frac{1}{6}bh^2 f_m + \sum_{i=1}^{m} A_{si} f_{si} j d_i \tag{1.8}$$

Equation (1.8) can be used to calculate the moment of resistance of the section up to the point where the first tensile crack occurs.

The first tension crack will occur when the maximum mortar tensile stress, f_m, reaches the flexural tensile strength of mortar, called modulus of rupture. In the absence of experimental results, the following empirical relation can be used to predict the modulus of rupture, f_r.

$$f_r = 0.62\sqrt{f'_c} \tag{1.9}$$

where f'_c is the compressive strength of mortar expressed in MPa.

In field applications, very rarely, the maximum stress will be less than the modulus of rupture. Hence, eq. (1.8) is not directly used to design dimensions of a structural member. However, the moment of resistance at first crack, M_{cr}, called the cracking moment is useful for the stiffness calculations in the post-cracking stage. The cracking moment M_{cr} can be calculated by substituting f_r for f_m in eq. (1.8). Because of the empirical nature of the stiffness calculations, explained subsequently, an extremely accurate evaluation of the cracking moment is not necessary. Instead, the cracking moment can be calculated using the following simplified equation

$$M_{cr} = \frac{1}{6}bh^2 f_r[1 + (n-1)V_{RL}] \tag{1.10}$$

where V_{RL} is the volume fraction of the reinforcement in the bending direction $(\sum_{i=1}^{m} A_{si}/bh)$ and n is the modular ratio E_R/E_m. In eq. (1.10), it is assumed that the wire meshes are uniformly and continuously distributed throughout the cross-section. Equation (1.10) can also be used to calculate the moment of resistance M for any given stress f_m by replacing f_r with f_m.

Using the classical bending theory and the simplified procedure, the equivalent stiffness of ferrocement in the precracking stage can be expressed as $E_m I_g$, where I_g is the moment of inertia of the uncracked cross-section determined by the equation:

$$I_g = \frac{1}{12}bh^3[1 + (n-1)V_{RL}] \tag{1.11}$$

In summary, in the precracking stage of loading, the moment capacity and the moment of inertia (needed for stiffness computations) can be calculated by using eqs. (1.10) and (1.11), respectively.

1.3.3.2 *Post-cracking stage.* The post-cracking stage starts with the occurrence of the first crack[45]. This stage extends up to the point where the extreme tension fibre of reinforcement starts yielding. The load–deflection behaviour in this stage of loading represents the behaviour of ferrocement in field

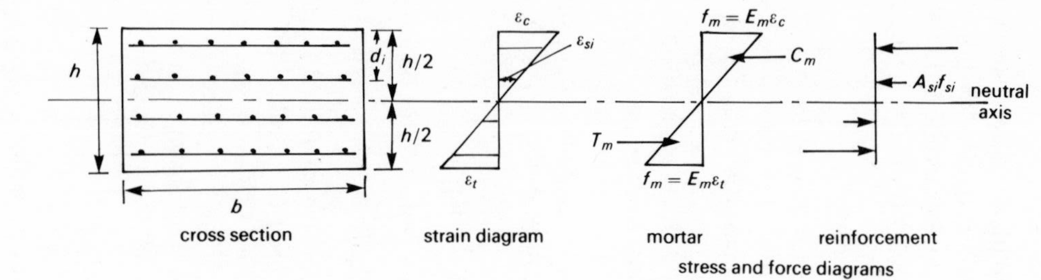

Figure 1.6 Distribution of strains, stresses and forces—uncracked section of the beam

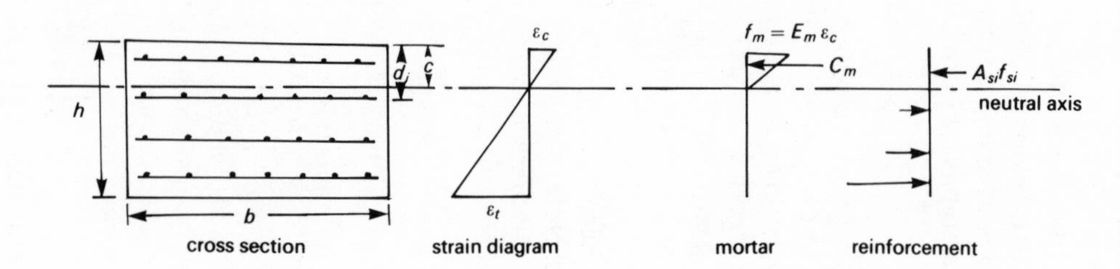

Figure 1.7 Distribution of strains, stresses and forces—cracked section of the beam

conditions, where almost all beams are cracked in the tension zone but the stress in the extreme tension fibre is well within the yield strength.

The moment resistance of the beam can be calculated by using the classical theory as in the case of precracking stage (Fig. 1.6) and eq. (1.7). However, since the section is cracked, the following important modifications have to be made to eq. (1.7).

After cracking the tensile force contribution of mortar is negligible compared to the contribution of reinforcement, and hence T_m can be assumed to be zero. The stresses and forces can be represented as shown in Fig. 1.7.

The depth of neutral axis can be calculated by using the force equilibrium equation:

$$\tfrac{1}{2}bcf_m = \sum_{i=1}^{m} A_{si}f_{si} \tag{1.12}$$

Since $f_m = E_m\varepsilon_c$, $f_{si} = E_R\varepsilon_{si}$, $(\varepsilon_c/\varepsilon_{si}) = (c/(d_i - c))$ and $n = E_R/E_m$, eq. (1.12) can be rewritten as:

$$\frac{bc^2}{2} = n\sum_{i=1}^{m} A_{si}(d_i - c) \tag{1.13}$$

The term $(d_i - c)$ becomes negative for reinforcement layers located in the compression zone. Note that in eqs (1.12) and (1.13) both steel and mortar are assumed to be in the linear elastic stage.

Using the moment equilibrium condition, the resisting moment, M can be expressed as:

$$M = \frac{bc^2}{3}f_m + \sum_{i=1}^{m} A_{si}f_{si}(d_i - c)^2 \tag{1.14}$$

Equation (1.14) can also be written in a form convenient to calculate the stresses in mortar and reinforcement as follows:

$$f_m = \frac{M}{I_{cr}}c \tag{1.15}$$

$$f_{si} = \frac{M}{I_{cr}}(d_i - c)n \tag{1.16}$$

where I_{cr} is the moment of inertia of the cracked section calculated using the equation

$$I_{cr} = \frac{bc^3}{3} + \sum_{i=1}^{m} nA_{si}(d_i - c)^2 \tag{1.17}$$

Equations (1.14) to (1.17) essentially represent all the equations needed for design calculations in the post-cracking, pre-yielding stage of loading.

In order to satisfy serviceability criteria, specified in terms of maximum permissible deflections and crack widths depending on the type of structure, one has to calculate both these parameters in addition to the moment capacity.

1.3.3.3 *Deflection calculations.* Using the principles of strength of materials, the maximum deflection, δ, of a typical beam can be obtained using the equation

$$\delta = \frac{kPL^r}{EI} \tag{1.18}$$

where k and r are constants which would depend on loading and support conditions, P is the load, L is the span and EI is the flexural rigidity.

Equation (1.18) can be used for ferrocement beams if a proper value is substituted for the flexural rigidity, EI. A typical ferrocement beam will have both cracked and uncracked regions. Uncracked sections of the beam will have a higher rigidity than the cracked section. The mortar in the tension zone located between cracks will contribute to the flexural rigidity. The decrease in rigidity in a certain portion of the beam would depend on the number and extent of penetration of the cracks. Since it is extremely difficult to consider all these variables, the following empirical approach adopted from American Concrete Institute Code[47] can be considered as sufficient for design calculations.

For ferrocement beams the flexural rigidity can be expressed as $E_m I_e$ where E_m is the Young's modulus of mortar (in MPa),

$$E_m = 4730\sqrt{f'_c} \tag{1.19}$$

and I_e is the effective moment of inertia

$$I_e = I_{cr} + \left(\frac{M_{cr}}{M_a}\right)^3 (I_g - I_{cr}) \leqslant I_g \tag{1.20}$$

where M_a is the maximum moment along the span of the beam and M_{cr}, I_g and I_{cr} can be calculated using eqs. (1.10), (1.11) and (1.17) respectively. Equation (1.19) is an empirical relation developed using a number of experimental investigations[47] for conventionally reinforced concrete.

For any given ferrocement beam the flexural rigidity can be calculated using Eqs. (1.19) and (1.20), which in turn can be used in eq. (1.18) to obtain the deflection.

The deflections and stresses calculated based on the properties of the transformed section, as explained above, were found to be in good agreement with the results obtained using non-linear computerized analysis based on the actual stress–strain relationships of mortar and wire meshes[33].

1.3.3.4 *Crack width determination—approximate method.* The study of the existing literature on cracking of ferrocement beams leads to the following observations[11,26,33,36,37,45] for the situation where the reinforcement was made up of wire meshes with square openings. Data available for specimens reinforced with other type meshes such as chicken-wire mesh are not sufficient to formulate a reasonably valid crack-width prediction equation.

The average crack width is primarily a function of the tensile strain in the extreme layer of mesh and does not depend on the other parameters such as clear cover of the tension reinforcement, found to be important in reinforced concrete. This may be because of the relatively small cover and high specific surface of reinforcement in ferrocement beams.

The transverse wires of the outermost layer of mesh are preferential locations for cracks.

The specific surface of reinforcement did not seem to have as strong an influence on the cracking behaviour in flexure as in tension. Nevertheless, the specific surface should be considered if one needs to predict the crack width more accurately[37].

Based on the above observations, the average crack width can be predicted using the following simple equation, in which the crack width is expressed as a function of strain at the extreme tension face and the spacing of transverse wires.

$$W_{av} = S\beta\varepsilon_s \tag{1.21}$$

where W_{av} = average crack width, S = mesh size or spacing centre-to-centre of wires, β = ratio of distance to the neutral axis from the extreme tensile fibre and from the outermost layer of steel, and ε_s = strain in the extreme tensile layer of mesh.

If h is the total thickness of the beam, d_{max} is the depth measured to the extreme tension layer and c is the depth of neutral axis calculated using elastic cracked section analysis, eq. (1.13), then

$$\beta = \frac{h - c}{d_{max} - c} \tag{1.22}$$

Using eq. (1.16), the stress in the extreme tension layer of wire mesh,

$$f_{s\max} = \frac{M}{I_{cr}}(d_{max} - c)n \tag{1.23}$$

If E_R is the Young's modulus of the wire mesh, the strain in the extreme tension layer wire mesh

$$\varepsilon_s = \frac{f_{s\max}}{E_R} \tag{1.24}$$

Hence, one can calculate the strain ε_s, and the factor β using the cracked section analysis which, in turn, can be used in eq. (1.21) to calculate the average crack width.

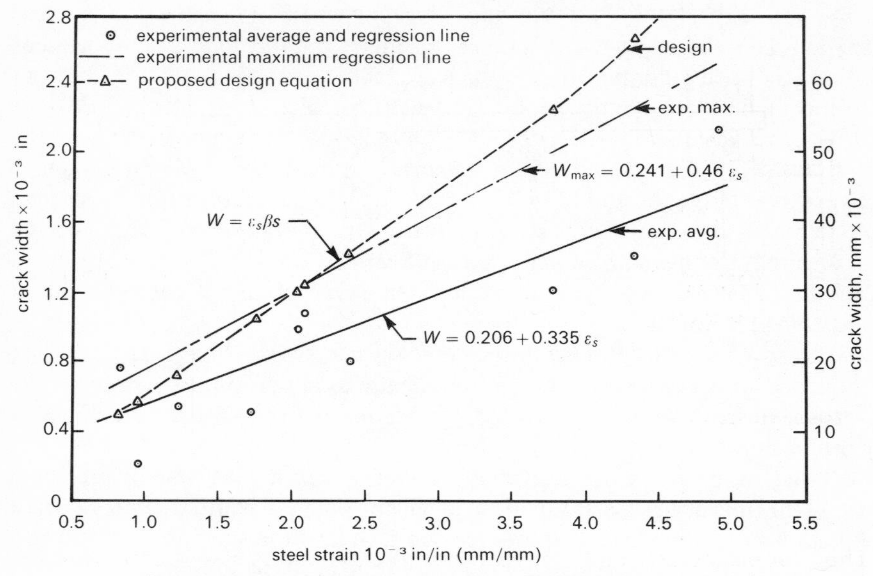

Figure 1.8 Crack width versus extreme steel fibre strain in flexure

The crack widths calculated using eq. (1.21) are compared with experimental results in Fig. 1.8. The experimental results were taken from ref. 33. The following are the details of the experiment. All the beam specimens were 457 mm long, 127 mm wide and 12.7 mm thick. The mortar consisted of rapid hardening cement (ASTM Type III), river sand passing through Sieve No. 16 (approximate maximum particle size of 1.8 mm), with a water–cement and sand–cement ratio by weight of 0.55 and 2.0, respectively. Three types of meshes with square openings were used as reinforcement: 12.7 mm woven, 12.7 mm welded and 6.35 mm woven. The wire diameters were 1.06 mm for 12.7 mm welded and 12.7 mm woven and 0.64 mm for 6.35 mm woven. For each type of mesh, three specimens each were made with two, four and six layers. The specimens were cast in vertical Plexiglas moulds to assure two smooth surfaces and to better control the locations of mesh layers. They were moist-cured for 7 days and left under the laboratory environment (21°C, 50% relative humidity) for 4 days before testing.

The beams were tested in an MTS (Materials Testing Systems) universal testing machine with a rate of stroke of 2.5 mm/min. A four-point loading arrangement with a 381 mm span and 127 mm constant bending moment zone was used (Fig. 1.9). The support system was carefully designed so as to avoid any axial forces on the beam. Crack-width measurements were made by stopping the loading at a selected number of deflection values. When the cracks were formed, they ran the full width of the specimens. The crack widths were measured along two longitudinal lines in the constant moment zone.

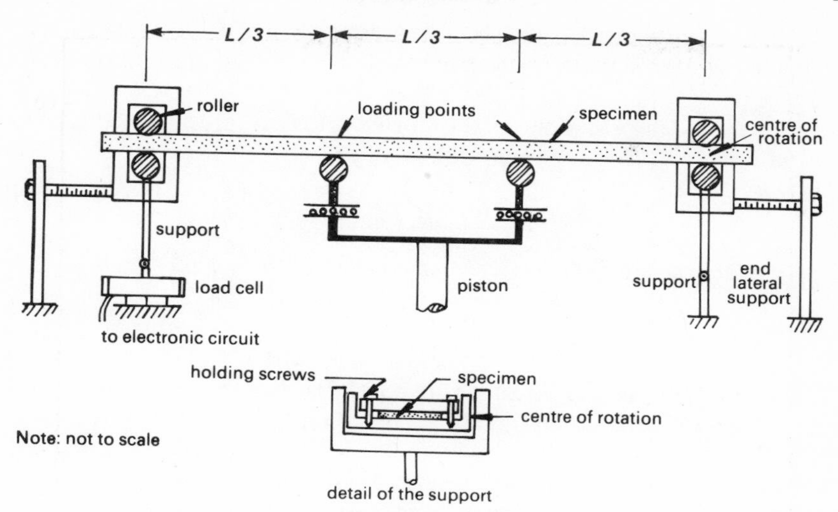

Figure 1.9 Schematic view of loading set-up for bending test

Thus, for each crack, and at each deflection, there were two width measurements. Crack measurements were made by an internally illuminated microscope equipped with a micrometer (50 × magnification), with a measuring accuracy of 0.001 27 mm.

In Fig. 1.8, it can be observed that eq. (1.21) provides a useful upper bound value for the crack widths. It is important to note that each point in Fig. 1.8 has a substantial weight since it represents an average of as many as 60 measurements. As an approximation, the crack width calculated using eq. (1.21) can be considered as maximum crack width. However, if one needs to calculate the crack widths more accurately, the following refinements can be made to eq. (1.21) using the results of ref. 37.

In eq. (1.21), the tension-stiffening effect of the mortar between cracks is neglected and the crack spacing is assumed to be equal to the transverse wire spacing. These two assumptions simplify the calculation of crack width. However, the tension-stiffening effect can be considered in the calculation if needed for accuracy. Also, a more accurate expression can be developed for predicting the crack spacing.

1.3.3.5 *Crack width determination—more accurate method.* Using the well-accepted Branson's[48] approach developed for reinforced concrete, an expression for the effective moment of inertia, I_{ec}, in the cracked region of the beam can be written as

$$I_{ec} = I_{cr} + \left(\frac{M_{cr}}{M_a}\right)^4 (I_g - I_{cr}) \tag{1.25}$$

Equation (1.25) is similar to eq. (1.20) but for the exponent 4 used for (M_{cr}/M_a).

Note that eq. (1.20) predicts the effective moment of inertia of the entire beam whereas eq. (1.25) predicts the effective moment of inertia of the cracked region of the beam.

If E_m is the Young's modulus of the mortar, the curvature of the beam in the cracked region, ψ, can be expressed as

$$\psi = \frac{1}{E_m I_{ec}}. \tag{1.26}$$

The equivalent strain in the extreme tension face of the beam ε_t can be written as:

$$\varepsilon_t = \psi(h - c) \tag{1.27}$$

ε_t in eq. (1.27) is analogous to '$\beta\varepsilon_s$' in eq. (1.21). The value of ε_t will be less than '$\beta\varepsilon_s$' since ε_s is based on I_{cr} and ε_t is based on I_{ec} and $1/I_{cr}$ is greater than $1/I_{ec}$.

By including the effect of specific surface of the wire meshes located in the tension zone, an improved expression for the crack spacing, a_c, can be written as[37]:

$$a_c = \left(\frac{\theta}{\eta}\frac{1}{S_{RL}}\right)\sqrt{\frac{\psi_y}{\psi}} \tag{1.28}$$

where ψ, the curvature of the beam $= \dfrac{M}{E_m I_{ec}}$

ψ_y, the curvature of the beam at yielding of steel $= \dfrac{f_y}{E_R(d_{\max} - c)}$

θ = ratio of average crack spacing to maximum crack spacing (same as in tension members)

η = the ratio of bond strength between the mesh and the mortar to the tensile strength of the mortar (same as in tension members), and

S_{RL} = specific surface of the reinforcement (same as in tension members)

Using the results of refs. 21 and 22, for specimens reinforced with wire meshes, θ/η can be taken as 1. Moreover, cracks very rarely form in between the transverse wires, since they are very closely spaced. Hence, eq. (1.28) can be written in the modified form

$$a_c = \left(\frac{1}{S_{RL}}\right)\sqrt{\frac{\psi}{\psi_y}} \geqslant \text{transverse wire spacing, } s \tag{1.29}$$

Using the crack spacing a_c calculated from eq. (1.29) and the extreme tension fibre strain ε_t calculated from eq. (1.27) an accurate expression for the average crack width W_{av} can be written as:

$$W_{av} = a_c\varepsilon_t \tag{1.30}$$

Assuming the variation of crack widths follow a normal distribution curve, and using the statistical variation of experimental results on crack widths, the maximum crack width W_{max} can be expressed as:

$$W_{max} = 1.5 W_{av} \tag{1.31}$$

Equations (1.30) and (1.31) were shown to predict the crack widths more accurately than eq. (1.21) in ref. 37.

1.3.3.6 *Fatigue loading*

(1) Fatigue failure of ferrocement flexural specimens is governed by the tensile fatigue properties of the mesh, just as for reinforced and prestressed concrete beams. Based on the regression analysis of the experimental data (Fig. 1.10) for ferrocement beams subjected to varying amplitudes of completely reversed loading of up to one million cycles and reinforced with varying volume of square welded and woven meshes, the following equation relating the fatigue failure with the stress in the steel was derived[39]:

$$f_{sr} = 1051 - 137 \log_{10} N_f \tag{1.32}$$

where f_{sr} = stress range in the outermost layer of steel in MPa calculated from the cracked elastic section analysis, and N_f = number of cycles to failure.

(2) The following relationship was found[39] to represent the data on deflections, and average or maximum crack width (Figs. 1.11 and 1.12)

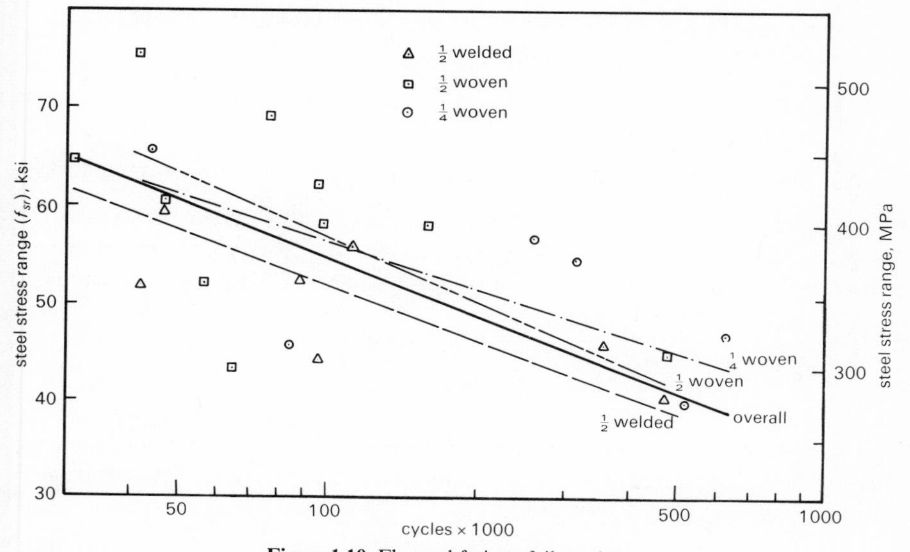

Figure 1.10 Flexural fatigue failure data

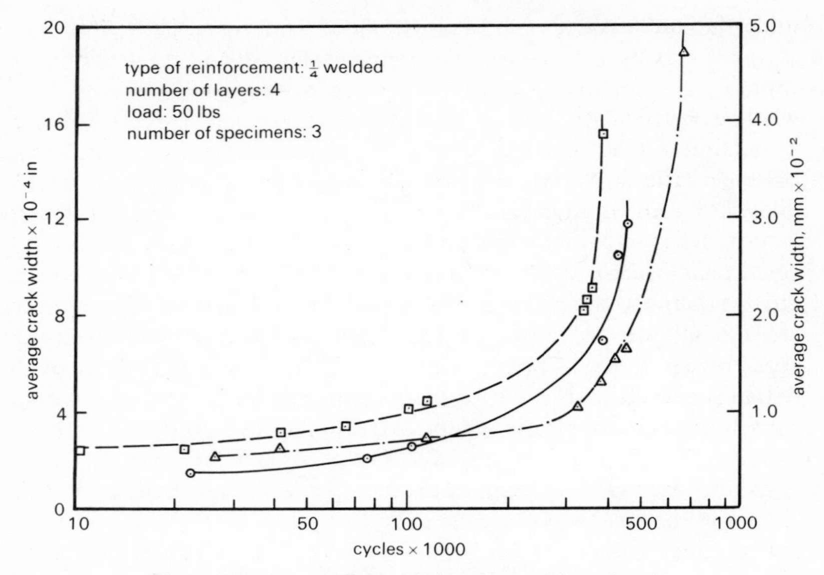

Figure 1.11 Average crack width versus number of cycles

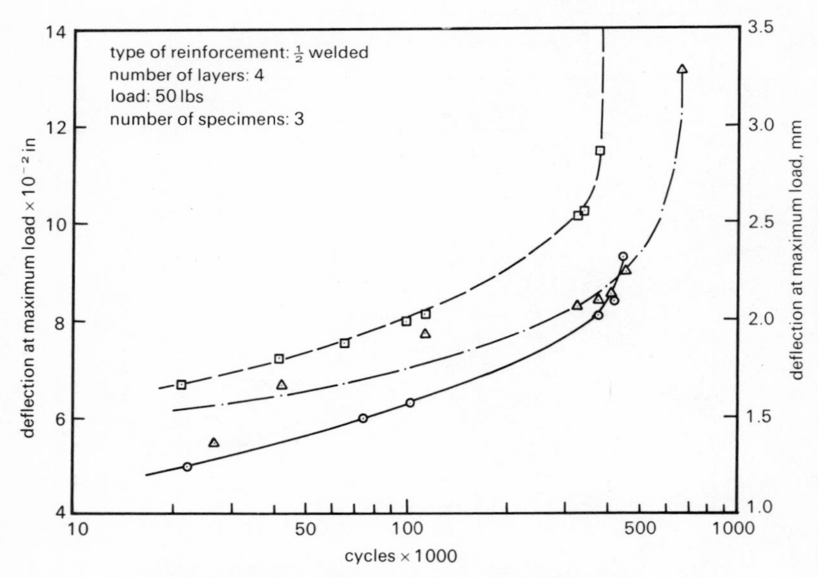

Figure 1.12 Maximum deflection versus number of cycles

under fatigue loading up to about 90% of the fatigue life:

$$Y = Ae^{Br} \tag{1.33}$$

where Y = deflection, average or maximum crack width; A = value of Y at maximum load under static test; e = base of Napierian logarithms; r = cycle ratio N/N_f, i.e., number of cycles N at which Y is calculated, divided by the number of cycles to failure, N_f; B = a parameter determined from the experimental data[39], and related to N_f. A constant value of $B = 0.667$ for deflection and $B = 1.67$ for crack width for N_f greater than 450 000 cycles was observed. For calculation of maximum crack width, the value of constant A can be determined by using eq. (1.21) or eq. (1.31). In order to use eq. (1.33), the number of cycles to failure is needed. This can be calculated for a given design situation by calculating the stress range in the extreme tensile mesh layer from eq. (1.32).

(3) A typical example of crack width increase with number of cycles is shown in Fig. 1.13 to illustrate the use of eq. (1.33). It can be seen that for lower value of stress range (higher value of N_f) the increase in crack width at a given number of cycles N due to fatigue loading is much less than for the higher value of the stress range (lower value of N_f).

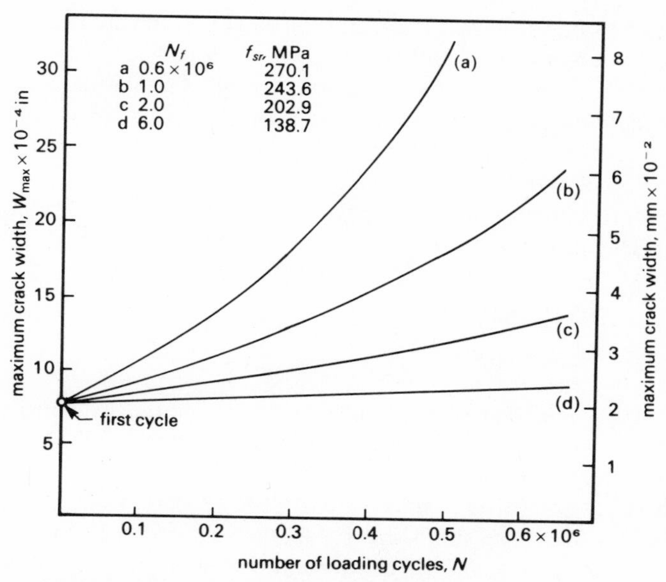

Figure 1.13 Typical crack width increase with loading cycles[21]

1.3.4 *Behaviour under impact loading*

Based on the very few investigations available in the published literature[43,44], the following observations can be made regarding the behaviour of ferrocement under impact loading.

For the same volume of reinforcement, increase in specific surface increases the impact strength.

Impact resistance increases almost linearly with the increase in volume fraction of reinforcement.

For equal value of reinforcement percentage, welded-wire meshes offer the highest impact resistance. In comparison, the specimens reinforced with chicken-wire meshes offer the lowest impact resistance. Woven mesh reinforcement provides impact strength higher than chicken-wire meshes but lower than welded-wire meshes.

As one could expect, different geometry of projectiles produce different types of failure patterns. It was found that: a spherical projectile causes a 'crater' type of failure, the cylindrical projectile causes a 'shear cut out of a plug' and the conical projectile causes a 'penetration' failure.

1.4 Recommendations for material selection and design

In this section, recommendations are made for the selection of materials, their composition and the design of ferrocement structures. These guidelines are based on the current state of knowledge, some of which was summarized in the preceding section. Additional guidelines especially in the areas of fabrication and quality control can be found in references 49 to 53.

No recommendations are made regarding impact strength, shear strength or creep behaviour of ferrocement since sufficient information is not available on these properties. Information related to these behaviours can be found in references 43, 44 and 54 to 59.

This section is divided into three parts dealing with: material requirements, design specifications and a numerical example illustrating the design calculations.

1.4.1 *Material requirements*

1.4.1.1 *Matrix.* The matrix used in ferrocement primarily consists of hydrated portland cement and inert aggregate. Pozzolans may be blended with the cement for special applications. Normally, the aggregate consists of well-graded sand passing through number 8 (ASTM)[60] sieve. If the reinforcement mesh openings are large, small-size gravel may be added to the sand.

Matrix represents about 95% or more of the ferrocement and has a great influence on the behaviour of the final product. Hence great care should be exercised in choosing the constituent materials, namely cement and aggregate, mixing and placing of the mortar. Chemical composition of the cement, the nature of the aggregate, aggregate–cement ratio and water–cement ratio are the major influencing parameters in determining the property of the mortar. Detailed studies on the various matrix mix proportion parameters can be found in references 12, 45 and 61 to 64.

1.4.1.2 *Cement.* The cement used should conform to ASTM C150 or an equivalent standard. Since manufacturers of cement have to follow certain standards stipulated by the government or code-writing bodies, it is not difficult to maintain the quality of cement. The cement should be fresh, of uniform consistency and free of lumps and foreign matter. Cement should be stored under dry conditions and for as short duration as possible.

The most commonly used cement type is called all-purpose cement and designated as Type I in ASTM[60]. Type II cement generates less heat during hydration and is also moderately resistant to sulphates. Type III is a rapid-hardening cement which acquires design strength more rapidly than Type I cement. Type IV is a low-heat cement used for mass concrete and is very little used for ferrocement. Type V is a sulphate-resisting cement used in structures exposed to sulphates.

The choice of the particular cement should depend on the service conditions. Service conditions can be classified as passive or active. Ground structures such as ferrocement silos, bins and water tanks can be considered as passive structures. For these structures, if the environment is not hostile such as a location near a sulphate factory, Type I cement can be used. For structures subjected to active service conditions such as boats and barges it might be necessary to specify sulphate-resisting cement because of the sulphates present in the sea water. If sulphate-resistant cement is not available, a rich mortar mix with normal cement can be used with proper and adequate surface coatings. Blended hydraulic cement can also be used. Blended cement should conform to the ASTM Standards[60] C595 Type IS, IS-A, IP or IP-A.

1.4.1.3 *Aggregates*[65–70]. The most common aggregate used in ferrocement is sand. Sand should comply with ASTM standard C33[60] (for fine aggregate) or an equivalent. It should be clean, inert and free of organic matter and deleterious substances and relatively free of silt and clay. Hard, strong and sharp silica sand provides the best results.

Sand should be uniformly graded. Table 1.2 can be used as a guideline for the grading. Normally the maximum particle size should be restricted to about 1 mm (0.04 inch). Grading of sand should be done to ensure high-density workable mortar mix.

Some parts of the world have aggregates which react with alkali in cement, which should be avoided. If in doubt, tests should be conducted to ensure the inert nature of the aggregate.

Lightweight sand can also be used for the ferrocement. The possible

Table 1.2 Guidelines for grading of sand

Sieve size U.S. standard square mesh[60]	*% passing by weight*
No. 8	100
No. 16	50–85
No. 3	25–60
No. 50	40–30
No. 100	2–10

lightweight aggregates are: volcanic ash, slag, diatomaceous earth, expanded shale fines, perlite, pumice, vermiculite, and inert alkali-resistant plastics. Most lightweight aggregates reduce the strength of the matrix and hence the corresponding adjustments should be made in the structural design.

Lightweight aggregate should conform to ASTM C330[60] or an equivalent.

1.4.1.4 *Water*. The mixing water should be fresh, clean and potable. The water should be free from organic matter, silt, oil, sugar, chloride, and acidic material. Salt water is not to be used. Normally drinking water can be used to produce acceptable results.

1.4.1.5 *Admixtures*. Admixtures used in the ferrocement serve one of the following four purposes; namely: (i) water reduction, increasing strength and reducing permeability, (ii) water-proofing, (iii) increased durability, and (iv) reduced reaction between the matrix and galvanized reinforcements[12,65,66].

Even though a number of water-reducing admixtures are available, the popular ones in recent times are called high-range water reducers or 'superplasticizers'. The use of water-reducing agents permits the use of more sand for same design strength which also results in fewer creep strains and less surface cracking.

If watertightness is extremely important, as in water- or liquid-retaining structures, waterproofing compounds can be used. In order to achieve watertightness, water/cement ratio should be kept below 0.4.

Pozzolans such as fly ash can be added to the cement to increase the durability. Up to a maximum of 30% of cement can be replaced with pozzolans without reducing the ultimate strength[45]. Air-entraining agents can be used to increase durability of the structures subjected to a large number of freeze–thaw cycles.

If galvanized reinforcing material is used, it was found that chromium trioxide added to the mix water reduces the reaction between the matrix and the reinforcements[12]. The recommended dosage, which depends on water/cement ratio, is approximately 300 parts per million by weight of mortar.

It is worthwhile to mention that if good judgement is exercised in choosing the material, mix design, placing and curing, quality matrix can be obtained without using any admixtures. In special situations such as a corrosive environment, proper additives or coatings must be used in order to ensure long and reliable serviceability.

If admixtures are used, they should conform to ASTM (or equivalent) standards[60] C260 (air-entraining admixtures), C494 (chemical admixtures), and C618 (fly ash and pozzolanic materials). If a new additive not covered in the specifications is used, tests should be conducted to obtain the properties of fresh and hardened mortar.

1.4.1.6 *Mix design.* The recommended mix proportions are: sand/cement ratio by weight 1.5 to 2.5, water/cement ratio by weight 0.35 to 0.5. Fineness modulus of sand, water/cement and sand/cement ratio should be carefully controlled to maintain the quality of the matrix.

The moisture content of the aggregate should be taken into account in the calculation of required water. Quantities of materials are normally to be determined by weight. However, volume method can also be used, provided enough care is taken to obtain the accurate bulk density. Allowance should also be made for bulking due to moisture.

The amount of water can be reduced if special admixtures are used. The mix should have as little water as possible. Normally the slump of fresh mortar should not exceed 50 mm (2 in). The 28-day compression strength of

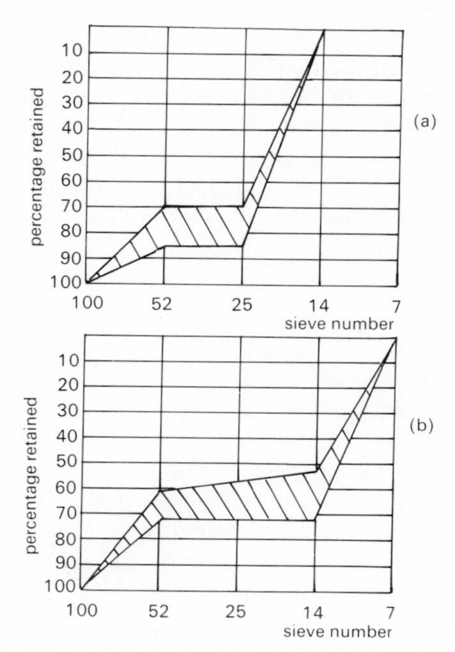

Figure 1.14 Optimal grading zone of sand[68]. (*a*) Maximum grain size, 1.2 mm; (*b*) maximum grain size 2.4 mm

75 mm × 150 mm (3 in × 6 in) moist-cured cylinders should be around 35 MPa (5000 psi) for most applications.

It is an established fact that minimum water should be used to obtain the maximum strength. For the given aggregates, workability of the mix determines the minimum water. In ferrocement the most common aggregate is sand. Hence, the properties of the sand have a major influence on the amount of water and hence the mix design. The following observations taken from reference 68 provide some guidelines for the designer.

1. The main parameters governing the water requirement of a mortar made with natural quartz sand are the maximum grain size, the fineness modulus and the grading composition. Improvement of the last named parameter makes for considerable reduction of the water requirement.
2. For a given fineness modulus, the minimum water requirement corresponds to maximum specific surface. Increase of the surface area is obtained by means of gap grading, with the medium fraction omitted; by this means the coarse grains are allowed maximum continuity and the amount of cement paste is reduced. For mixes conventionally used for ferrocement (maximum size 2.4 and 1.2 mm, consistency dry to plastic) no segregation resulted from the use of gap grading.
3. To each maximum size, there corresponds an optimum fineness modulus: 2.9–3.0 for 2.4 mm, 2.5 for 1.2 mm. The acceptable deviation interval of the fineness modulus is a function of the tolerance of the w/c ratio. (For the recommended grading zones in Fig. 1.14, this tolerance does not exceed 0.1.)
4. In sand with maximum size 0.6 mm, the presence of fraction No. 100 makes for considerable increase of the water requirement. The optimum composition comprises fraction No. 52 and fines passing sieve No. 100. The best results were obtained for a mix in which the total absolute volume of cement and fines was about 300 cm^3 per liter of mortar. A change in the amount of cement must be accompanied by a corresponding change in that of the fines. Neither the size of the fines nor their type had any effect.
5. Of the three sand sizes studied, the coarse variants (2.4 mm and 1.2 mm) were clearly superior in terms of the water requirement. Accordingly, fine sand should be avoided in ferrocement mixes, and the sand composition should preferably consist of non-consecutive fractions.
6. Increase of the cement or water content of the mix results in increased separation of the sand particles, hence the importance of the sand composition, especially in lean mixes and stiff consistencies.

If coarse aggregate is used in the mix, trial mixes should be made to ensure the workability and strength requirements.

1.4.1.7 *Reinforcement*. The reinforcement should be clean and free from all loose materials such as dust, loose rust, coatings of paint, oil or similar substances.

The most commonly used reinforcement in ferrocement is wire mesh with closely-spaced wires. For the purpose of special applications or economy; expanded metal, welded-wire fabric, reinforcing wires or rods, prestressing tendons, discontinuous steel fibres, glass fibres and organic fibres are also being used.

Wire mesh reinforcement should conform to ASTM standard[77] A-185 or an equivalent. Wire meshes have either hexagonal or square openings (Fig. 1.15). Meshes with hexagonal openings are sometimes referred to as chicken-wire

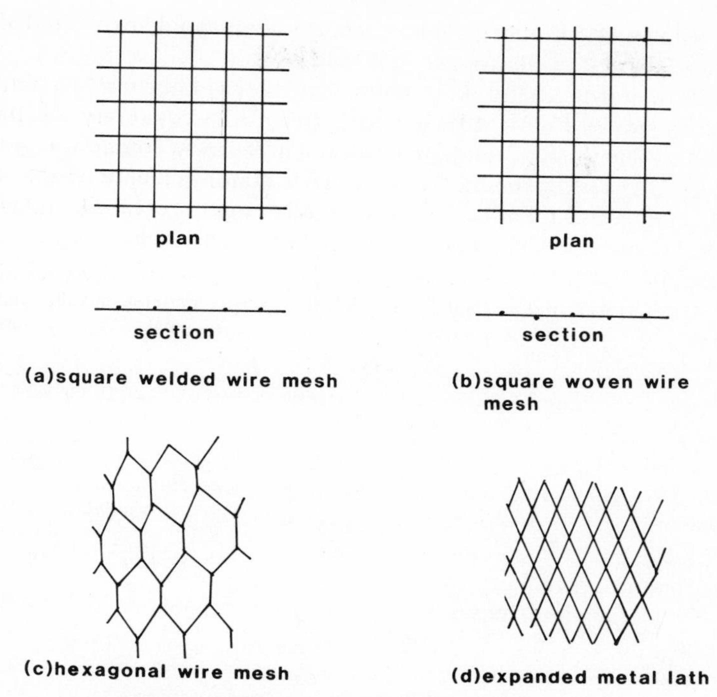

Figure 1.15 Different types of reinforcing materials

meshes. Meshes with hexagonal-shaped openings are not structurally as efficient as the meshes with square openings because the wires are not oriented in principal (maximum) stress directions. But they are very flexible and can be used in very thin sections.

Meshes with square openings are available either in in the form of welded-wire mesh or in the woven form. Welded-wire meshes have a higher Young's modulus and hence higher stiffness and less cracking in the early stages of loading. On the other hand woven-wire meshes are a little more flexible and easy to work with than the welded meshes. In addition, welding anneals the wire and limits the tensile strength[70]. Currently most designers recommend square woven mesh with 1 mm (19 gauge) diameter wires or 1.6 mm (16 gauge) diameter wires spaced about 13 mm (0.5 in.) apart.

Wire meshes are also available in the galvanized form. Galvanizing, like welding, reduces the tensile strength. Galvanized meshes when used with regular reinforcing bars may react with them to produce hydrogen bubbles. This reaction can be passivated by adding chromium trioxide to the mixing water at the rate of 200 to 300 parts per million[12].

Welded wire fabric differs from welded-wire mesh in the size and spacing of the wires. Welded-wire fabric normally contains larger-diameter wires spaced at 25 mm (1 in.) or more.

The welded-wire fabric should be used in combination with wire meshes where control of cracking is essential. The fabric should conform to ASTM standards[77] A-496 and A-497 or an equivalent. The minimum yielding strength of the wire should be 450 MPa (65 000 psi) for smooth wires and 480 MPa (70 000 psi) for deformed wires. The wire diameter should be less than 13 mm (0.5 in.) except in cases of very thick plates. A popular welded-wire fabric is a square mat made of 2 mm (14 gauge) wires spaced at 25 mm (1 in.).

Expanded-metal lath is formed by slitting thin gauge sheets and expanding them in a direction perpendicular to the slits (Fig. 1.15). The use of expanded metal was first studied by Collen in 1959[71]. Considerable further research findings were reported by Byrne and Wright[73], Johnston and Mowat[20], and Iorns[46]. The general conclusions were [20,46]:

Expanded metal and welded-wire mesh offer approximately equal strength in their normal orientation.

Expanded metal results in a stiffer composite when compared with welded mesh. This aspect reduces the crack widths at the earlier stage of loading.

Expanded metal provides better impact resistance and better crack control.

Expanded metal is suitable for both hulls and integral tanks if proper construction procedures are used.

Despite all the aforementioned advantages, expanded metal is not suitable for some applications. It is not flexible and hence cannot be used in construction involving sharp curves. However, in some countries expanded metal is cost-effective compared to wire reinforcements and should be considered as a possible design alternative.

Reinforcing rods and *prestressing wires* are used in combination with wire meshes for relatively thick ferrocement elements. Reinforcing bars should conform to ASTM standards[77] A-615 and A-616 or an equivalent. Prestressing wires should confirm to ASTM standards[77] A-421 and A-416. Under normal conditions reinforcing bars should be Grade 60 steel with a yield strength of 414 MPa (60 000 psi) and a tensile strength of 620 MPa (90 000 psi). The Young's modulus should be 200 GPa (30×10^6 psi). The prestressing wires should have an ultimate strength of either 1725 MPa (250 000 psi) or 1863 MPa (270 000 psi) and have a Young's modulus of 186.3 GPa (27×10^6 psi).

The minimum possible size of bar should be used in order to realize the effect of wire meshes and hence the composite effect.

Addition of *fibres* to ferrocement seem to enhance the properties considerably[73]. The steel wires seem to assist in distributing cracks and hence allow the use of much heavier woven- or loomed-wire meshes. Impact resistance was found to be increased. The various forms of fibres, their properties and their use are dealt with in detail in references 74 and 75.

1.4.1.8 *Testing for mechanical properties.* Tests should be performed to obtain the mechanical properties of the two main constituent materials namely, mortar and wire mesh. Mortar specimens should be tested to obtain the Young's modulus, compressive strength, modulus of rupture and the split tensile strength of the matrix. Tests on wire meshes should be done to obtain the Young's modulus, yield strength and the tensile strength. For certain applications, additional tests may be required to determine properties under special conditions such as impact loading.

The *compressive strength* of the mortar matrix can be determined using cubes, prisms, or cylinders. ASTM standard[60] C-496 or an equivalent should be followed. If Young's modulus is needed then the stresses and strains should be measured at least in the initial loading stage.

The two major parameters required for design, namely the compressive strength and the Young's modulus, can be obtained from the stress–strain curve (Fig. 1.16). At least three different methods are available to estimate the Young's modulus. The secant modulus method provides satisfactory results for most design applications. In this method, the Young's modulus, E_m is expressed as

$$E_m = \frac{\text{stress at 0.4 compressive strength}}{\text{strain at 0.4 compressive strength}} \tag{1.34}$$

Note that the value of E_m can also be estimated using an empirical equation, eq. (1.19), presented in the section on mechanical properties.

Two types of tests can be used to measure the *tensile strength* of mortar. One measure of tensile strength can be obtained using split cylinder test, ASTM C-496[60]. In this method, a cylinder 75 mm × 150 mm (3 in × 6 in) is subjected to a compression force across the diameter of the cylinder using two steel platens.

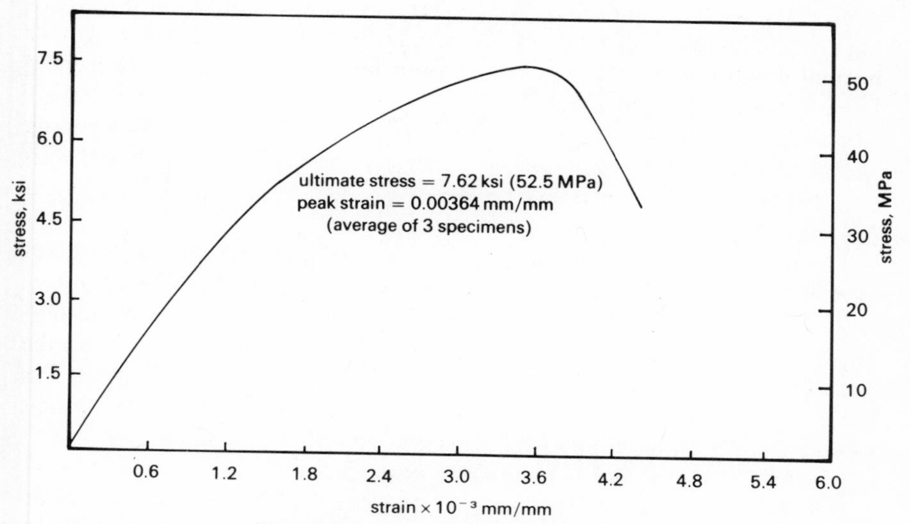

Figure 1.16 Stress–strain curve of mortar[/6]

If the failure load is P and l and d are the length and diameter of the cylinder, the split tensile strength of mortar f_{tm} can be calculated using the equation[60]

$$f_{tm} = \frac{2P}{\pi l d} \tag{1.35}$$

In the absence of experimental results, the following empirical equation can be used to estimate the tensile strength of normal-weight mortar. Similar relations are available for lightweight mortar in reference 47.

$$f_{tm} = 0.55\sqrt{f'_c} \tag{1.36}$$

The flexural tensile strength, i.e. the modulus of rupture, can be obtained by testing plain mortar beams under either third-point loading or centre-point loading. Certain guidelines have to be followed regarding the dimensions of the beams[60]. The modulus of rupture, f_r, can be obtained using the equation

$$f_r = \frac{M}{\frac{1}{6}bh^2} \tag{1.37}$$

where M is the maximum or failure moment, b is the breadth of the beam and h is the total thickness of the beam.

If P is the load at failure, and l is the span, then $M = Pl/4$ for centre-point

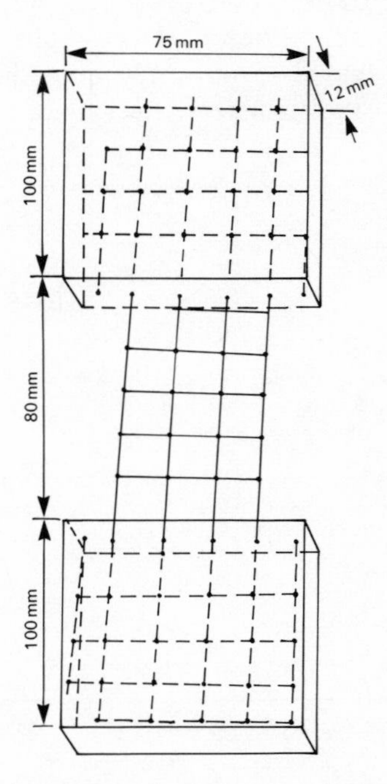

Figure 1.17 A typical tension test specimen for wire meshes[76]

loading and $M = Pl/6$ for third-point loading. Extensive details about this testing can be found in ASTM[60] C-293 and ASTM[60] C-683.

The modulus of rupture, f_r, can also be estimated using an empirical relation presented in eq. (1.9).

Reinforcement: only wire meshes are covered in this section. Other reinforcing materials such as rebars and prestressing wires, if used, should be tested using relevant ASTM standards[77].

Wire meshes used in ferrocement, normally consist of closely-spaced thin wires. Hence special type specimen preparation is necessary to determine the Young's modulus and the tensile strength of the wire meshes. Figure 1.17 shows[33] a typical tensile test specimen, for a square mesh with wires spaced at 13 mm[76] (0.5 in). A 50-mm (2 in) strip of mesh consisting of 4 wires is embedded in mortar on both sides, leaving a free mesh length of 75 mm (3 in). The mesh has more width (about $2\frac{3}{4}''$, 67 mm) inside the mortar. The mortar pads facilitate the easy gripping of dial gauges to measure strains. The mortar pads should be provided with extra mesh reinforcement, to prevent failure in the mortar portion of the specimen.

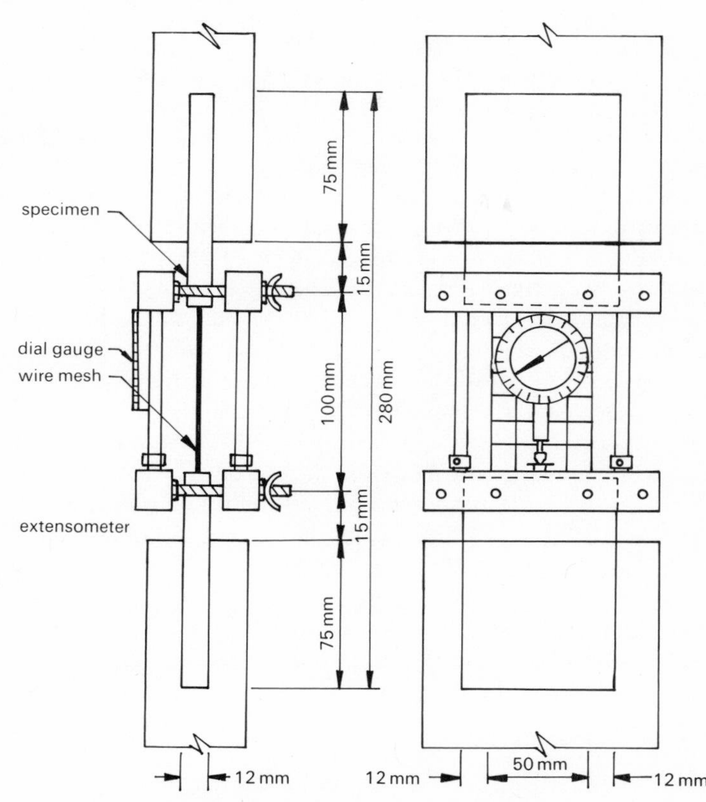

Figure 1.18 Tensile test set-up for wire meshes[76]

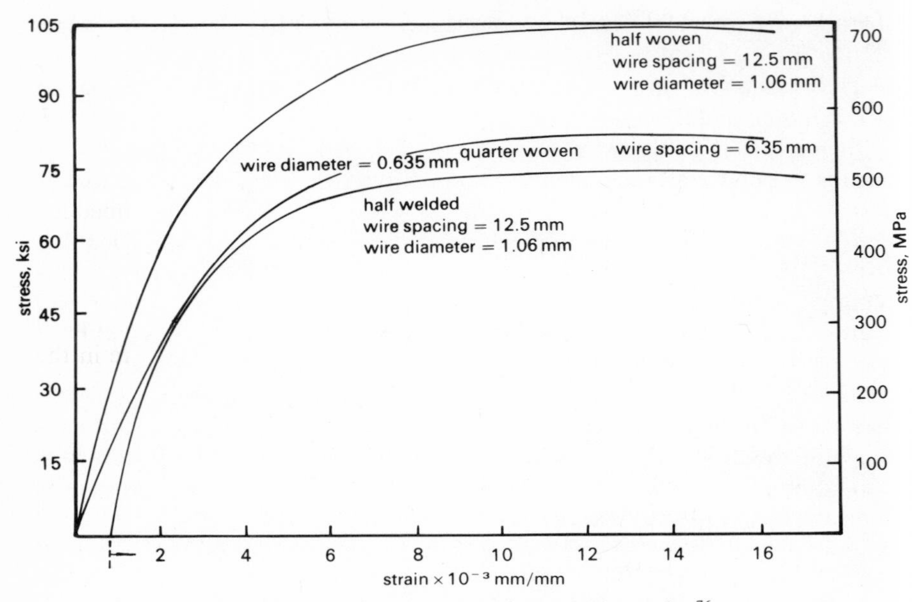

Figure 1.19 Stress–strain curves for steel wire meshes[76]

Figure 1.18, taken from ref. 33, shows the details of the dial gauge set-up. The dial gauge can be gripped to the mortar pads. The grip points should be as close as possible to the wire mesh in order to measure the strains accurately.

The stresses at various loading stages can be calculated using the loads recorded and the area of cross-section of the wires. The corresponding strains can be calculated using the dial gauge readings and the gauge length measured between the gripping points. Using various stress and strain values the stress–strain can be plotted. Figure 1.19 shows some typical stress–strain curves.

The initial slope of the stress–strain curve represents the Young's modulus of the wire reinforcement, E_R. It can be observed from Fig. 1.19 that there is no well-defined yield point. Hence the yield strength f_y should be determined using one of the following two methods.

In the first, called the 2% offset method, a straight line is drawn parallel to the initial slope starting at a strain of 0.002 mm/mm at the abscissa. The stress at which this line intersects the stress–strain curve is taken as the yield strength, f_y.

In the second method, the yield strength f_y is taken as the stress corresponding to a strain of 0.0035 mm/mm.

1.4.2 *Design recommendations*

(1) The allowable tensile stress in the steel reinforcement may be generally taken as $0.60 f_y$ where f_y is the yield strength measured at 0.0035 strain;

Table 1.3 Recommended values of yield strength, f_y and effective modulus, E_R, for steel wire meshes and reinforcing bars

	Woven square mesh	*Welded square mesh*	*Hexagonal mesh*	*Expanded metal lath*	*Longitudinal bars*
f_y, MPa (ksi)	450 (65)	450 (65)	310 (45)	380 (55)	410 (60)
E_{RL}, GPa (10^3 ksi)	140 (20)	200 (29)	100 (15)	140 (20)	200 (29)
E_{RT}, GPa (10^3 ksi)	160 (24)	200 (29)	70 (10)	70 (10)	— —

E_{RL} = modulus in the longitudinal direction
E_{RT} = modulus in the transverse direction

however, for water-retaining and sanitary structures it is preferable to limit the tensile stress to 200 MPa ($\simeq$ 30 ksi) unless crack-width measurements on a test model indicate that a higher stress would not impair performance. The above values hold, provided the pitch of the weave of the mesh system is moderate in order to ensure an adequate modulus.

The values of f_y and E_R presented in Table 1.3 can be used if the reinforcement conforms to the ASTM specifications[77] or an equivalent.

(2) The allowable compressive stress in the composite may be taken as 0.45 f'_c where f'_c is the specified compressive strength of the mortar measured from tests of 75 mm × 150 mm (3 in × 6 in) cylinders.

(3) The total volume fraction of reinforcement, V_R, in both directions, shall not be less than 1.8%. The total specific surface of reinforcement, S_R, in both directions, shall not be less than 0.8 cm^2/cm^3 (2 in^2/in^3). About twice these values are generally recommended (and more for water-retaining structures). In computing the specific surface of reinforcement the skeletal steel may be disregarded, while it may be considered in computing V_R.

(4) The recommended average net cover of the reinforcement is about 2 mm ($\frac{1}{12}$ in). However, a lesser value can be used provided the reinforcement is galvanized, the surface protected by painting and the crack width limited to a low value. It is also recommended that for thicknesses larger than 12 mm (0.5 in) the net cover shall not exceed $\frac{1}{5}$ of thickness t nor 5 mm ($\frac{3}{16}$ in) in order to ensure proper distribution of the mesh throughout the thickness.

(5) In order to predict the behaviour of ferrocement under service load conditions, an elastic analysis similar to that of reinforced concrete 'Working Stress Design' method is acceptable provided the modulus of the steel mesh system (which may be different from the modulus of the steel wire) is considered. Ultimate load can also be predicted for flexural

members by analysing ferrocement as a reinforced concrete member (or column) using the ACI ultimate strength design method[47]. For tensile members the ultimate load can be approximated by the load-carrying capacity of the mesh reinforcement alone in the direction of loading.

(6) It is recommended that the maximum value of crack width be less than 0.10 mm (0.004 in) for non-corrosive environment and 0.05 mm (0.002 in) for corrosive environment and/or water-retaining structures.

(7) The maximum crack width for flexural members can be predicted as a first approximation using eq. (1.21) or eq. (1.31).

(8) For ferrocement structures to sustain a minimum life of two million cycles the stress–range in the steel must be limited to $f_{sr} = 200$ MPa ($\simeq 30$ ksi). A value of $f_{sr} = 250$ MPa ($\simeq 36$ ksi) can be used for one million cycles and 380 MPa (55 ksi) for 100 000 cycles. For a given ferrocement material (without skeletal reinforcement) of thickness t the recommended mesh openings should not be larger than t. Furthermore, the number of layers of mesh, m should be such that:

$$m \geqslant 1.6\,t$$

where t = thickness in cm.

(9) If skeletal reinforcement is used, it is recommended that the skeletal reinforcement does not occupy more than 50% of the thickness of the ferrocement material. For this case use $m \geqslant 1.6\,t'$ where t' is the thickness in which meshes are distributed.

1.4.3 *Design example*

A cross-section of a ferrocement structure is shown in Fig. 1.20. It is required to calculate the adequacy of this section under the following service conditions: (a) a tensile force of 75 kN; (b) a compressive force of 150 kN or (c) a bending moment of 225 N-m. (Note that the calculations at ultimate capacity are straightforward and hence not demonstrated here.)

The reinforcement consists of 6 layers of welded-wire meshes distributed uniformly across the thickness. The wire spacing is 6.35 mm ($\frac{1}{4}$in) and the diameter is 0.635 mm. Given: the compressive strength of mortar, $f'_c = 35$ MPa, and the yield strength of reinforcement, $f_y = 450$ MPa.

Check the adequacy of the section in: (a) compression, (b) tension and (c) bending.

Solution. The stress resultants (forces) are specified for 1 m width. Hence the section to be analysed is 1 m wide and 12 mm thick. The specified wire spacing is 6.35 mm ($\frac{1}{4}$ in). This results in 158 longitudinal wires for each layer of wire mesh. Since the wire diameter is 0.635 mm, area of each layer of reinforcement $A_{si} = 50.04\ \text{mm}^2$.

In Fig. 1.20, which shows the cross-section and the stress distributions, note that:

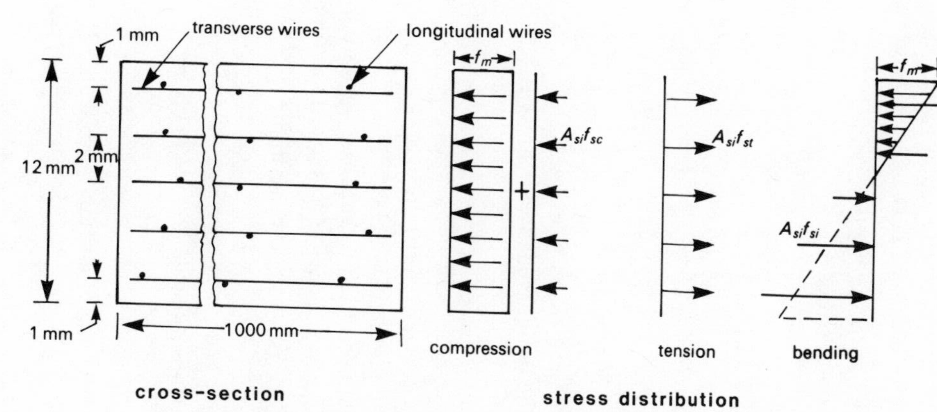

Figure 1.20 Cross-section and stress distribution (design example)

the stresses are the same for all the 6 layers of wire meshes for the compression and tension loading
for the tension loading, there is no mortar contribution to the strength of the composite member
for flexural loading, the stress distribution is linear across the thickness of the beam.

(a) Compression loading

Maximum compression force = 150 kN/m

If the contribution of wire meshes is neglected, maximum compressive stress

$$= \frac{150 \times 1000}{bh} \text{ MPa}$$

$$= \frac{150 \times 1000}{1000 \times 12} \text{ MPa}$$

$$= 12.5 \text{ MPa}$$

$$A_c = 1000 \times 12[1 + (7.15 - 1) \times 0.025]$$
$$= 13\,845 \text{ mm}^2$$

If the contribution of the wire meshes is considered in the design, maximum stress $= \frac{150 \times 1000}{13\,845}$ MPa

$$= 10.83 \text{ MPa}$$

Hence, if one considers the contribution of steel, there is a reduction of 1.7 MPa in the maximum stress.

The author's recommendation is that the contribution of steel need be considered only for very heavily reinforced sections or when the maximum stress calculated without considering steel exceeds the permissible stress.

(b) Tension loading

Maximum tensile force = 75 kN/m

Using eq. (1.36)

tensile strength of mortar, $f_{tm} = 0.55\sqrt{35}$
$= 3.25$ MPa

tensile strength of the uncracked cross-section $= A_c f_{tm}$

$$= 13\,845 \times 3.25$$
$$= 45 \text{ kN}$$
$$< 75 \text{ kN}$$

Hence the section cracks under the service loads.

As explained in section 1.3.2, once the mortar cracks, conservative upper limit values can be obtained for both stress and crack widths by assuming that the entire load is taken by the reinforcement.

$$\text{Total area of reinforcement} = 6 \times 50.04 = 300.24\,\text{mm}^2$$

$$\text{Maximum steel stress} = \frac{75\,000}{300.24} = 249.8\,\text{MPa}$$

$$\begin{aligned}\text{Permissible tensile stress}&\\ \text{(Section 1.4.2)}\quad &= 0.6f_y\\ &= 0.6 \times 450 = 270\,\text{MPa}\end{aligned}$$

Maximum stress is less than permissible stress. Hence design is satisfactory for tensile strength.

In order to calculate the maximum crack width, we need the actual steel stress and the specific surface, eqs. (1.3), (1.5) and (1.6).

$$\text{Steel stress, } f_s = 249.8\,\text{MPa}$$

$$\begin{aligned}\text{Specific surface, } S_{RL} &= \frac{\text{Perimeter of one wire}}{\text{area surrounding one wire}}\\ &= \frac{3.14 \times 0.635}{6.35 \times 2} = 0.157\,\text{mm}^{-1}\\ &= 1.57\,\text{cm}^{-1}\end{aligned}$$

Note that this equation is valid only for cases where the meshes are uniformly distributed across the thickness of the section.

$$345S_{RL} = 345 \times 1.57 = 541.7 > f_s = 249.8\,\text{MPa}$$

Using eq. (1.5),

$$\begin{aligned}\text{the maximum crack width, } W_{\max} &= \frac{3500}{E_R}\,\text{mm}\\ &= \frac{3\,500}{200\,000} = 0.00175\,\text{mm}\end{aligned}$$

This crack width is much less than the recommended maximum of 0.05 mm (section 1.4.2), hence the section is adequate.

(c) Flexural loading

$$\text{Maximum external moment} = 225\,\text{N-m}$$

Using eq. (1.9),

$$\begin{aligned}\text{Modulus of rupture } f_r &= 0.62f_c'\\ &= 0.62\sqrt{35} = 3.67\,\text{MPa}\end{aligned}$$

Using eq. (1.10),

$$\text{Cracking moment } M_{cr} = \tfrac{1}{6}bh^2 f_r[1 + (n-1)V_{RL}]$$
$$= \tfrac{1}{6} \times 1000 \times 12^2 \times 3.67[1 + (7.15 - 1) \times 0.025]$$
$$= 101.6 \text{ N-m}$$

External moment is greater than cracking moment and hence the cracked section analysis should be performed.

Using eq. (1.13), the depth of neutral axis c, for the cracked section can be calculated from (Fig. 1.23):

$$\frac{bc^2}{2} = n \sum_{i=1}^{m} A_{si}(d_i - c)$$

or

$$\frac{1000\,c^2}{2} = 7.15 \times 50.04[(11 - c) + (9 - c) + (7 - c) + (5 - c) + (3 - c) + (1 - c)]$$

or

$$c^2 + 4.29\,c - 18.60 = 0$$

or

$$c = 3.36 \text{ mm}$$

Using eq. (1.17),

the moment of inertia of the cracked section,

$$I_{cr} = \frac{1000 \times 3.36^3}{3} + 7.15 \times 50.04[(11 - 3.36)^2 + (9 - 3.36)^2 + (7 - 3.36)^2 + (5 - 3.36)^2 + (3 - 3.36)^2 + (1 - 3.36)^2]$$
$$= (12\,664 + 39\,841) \text{ mm}^4$$
$$= 52\,485 \text{ mm}^4$$

Using eq. (1.15),

$$\text{the maximum stress in concrete} = \frac{225 \times 1000}{52\,485} \times 3.36$$
$$= 14.4 \text{ MPa}$$
$$< 0.45 f'_c = 15.75 \text{ MPa}$$

Using eq. (1.16),

$$\text{the maximum stress in the extreme tension layer} = \frac{225 \times 1000}{52\,485}(11 - 3.36) \times 7.15)$$
$$= 234.2 \text{ MPa}$$
$$< 0.6 f_y = 270 \text{ MPa}$$

Using the simplified approach, the maximum crack width (eq. 1.21);

$$W_{max} = S\beta\varepsilon_s$$

$$S = \text{wire spacing} = 6.35\,\text{mm}$$

$$\beta = \frac{h-c}{d_{max}-c} = \frac{12-3.36}{11-3.36} = 1.131$$

$$\varepsilon_s = \frac{f_{s\,max}}{E_R} = \frac{234.2}{200\,000} = 0.001\,17\,\text{mm/mm}$$

Hence, $W_{max} = 6.35 \times 1.131 \times 0.001\,17 = 0.0084$ mm.

This maximum crack width is much less than the permissible maximum of 0.05 mm. Note that the accurate evaluation of crack width is not necessary for this case.

The section is adequate for flexural loading.

1.5 Applications

The use of ferrocement construction has been widespread only in the last 20 years[78–94]. The state-of-the-art of the application of this new construction material is still in its infancy. The long-term experience with ferrocement structures has not been sufficiently accumulated and analysed to assess the success of the structures already built. The main applications that have been made of ferrocement construction can be classified into three categories: boats, silos and tanks, and roofs.

The construction of ferrocement can be divided into four phases: (1) fabricating the skeletal framing system; (2) applying rods and mesh; (3) plastering; and (4) curing. Note that special skills are required for phases 1 and 3 while phase 2 is very labour-intensive, which may be a shortcoming for industrially developed countries but an advantage for countries where unskilled labour is relatively abundant. Laminating techniques similar to those developed by Fibersteel for marine structures can reduce the labour cost[78]. Experience has shown that the quality of mortar and its application is the most critical phase. Mortar can be applied by hand or by shotcreting. Note that since no formwork is required as in conventional reinforced concrete construction, ferrocement is especially suitable for structures with curved surfaces such as shells, and free-form shapes[92,93].

Ferrocement has very high tensile-strength-to-weight ratio and a superior cracking behaviour. This means that thin ferrocement structures can be relatively light and watertight. Boats, barges, mobile homes or other portable structures must be light in weight and impact-resistant. For these types of structures, ferrocement may be an attractive material. Building components which are seldom moved place only a moderate premium on light weight. Thus, even though ferrocement is more efficient, pound for pound, it is cheaper to build heavier structures in conventionally reinforced concrete. This is especially true in developed countries where high material costs and the more labour-intensive nature of ferrocement limit its use to only specialized applications. These include geodesic domes, wind tunnels, roof shells, mobile homes, modular housing, tanks and swimming pools. Even for these applications, in developed countries, the ready availability of mobile cranes and efficient systems for precasting, prestressing, and erecting reinforced concrete structures of any size make widespread use of ferrocement not very likely.

While ferrocement construction may not be very cost-effective compared to conventional concrete construction for many applications, it can compete favourably with fiberglass laminates or steel. Two recent feasibility studies have shown ferrocement to be cheaper than steel for the construction of wind tunnels[79] and than steel or fiberglass for construction of tanks for storing hot water[80].

1.5.1 *Boats*

Ferrocement boats have been now built in almost every country in the world. Some idea of the extent of ferrocement boat construction is given in Table 1.4[81] where a list of boat construction in the countries of the Asian/Pacific region is given. The People's Republic of China appears to be the only country

Table 1.4 Ferrocement boats built in asian/pacific region (reference 81)

Country	*Number of boats built*	*Size of boats (m)*	*Type of boat*	*Number of ferrocement boat building yards*
Bangladesh	3	10–14	Transport/fishing	1
China	2000 (est.)	12–15	Transport	30 (est.)
Hong Kong	4	15–27	Fishing	
India	6	5–11	Fishing	1
Indonesia				
Japan	10 (est.)			
Korea	11	10–25	Fishing	1
Malaysia	1			1
Pakistan	2			
Philippines	2			
Singapore	3		Transport/pleasure	
Sri Lanka	10	7–12	Fishing	1
Thailand	30	5–24	Pleasure, fishing transport	2
Vietnam, North	unknown			
Vietnam, South	50	7–20	Transport	unknown
Australia				
Fiji	11	10–15	Transport/fishing	1
New Zealand	500 (est.)	8–20	Pleasure, fishing tug boat	
Solomon Islands	3		Fishing	
Western Samoa	1	15	Fishing	

where ferrocement boats have been introduced on a large scale. In other countries ferrocement occupies only a fraction of a percent of the total boat building market.

Ferrocement boat construction has been found attractive for many industrially developing countries[92,93] because: (1) its basic raw materials are available in most countries; (2) it can be fabricated into almost any shape, traditional design can be reproduced and often improved; (3) it is more durable than most woods and cheaper than imported steel; (4) the skills for ferrocement construction can be easily acquired; (5) ferrocement construction is less capital-intensive and more labour-intensive; and (6) except for sophisticated and highly stressed design such as those in deep-water vessels, a trained supervisor can achieve the requisite amount of quality control using fairly unskilled labour.

Ferrocement is relatively heavy material compared with wood and fibre-reinforced plastic. Most wooden boats below 10 m in length are built with a plank thickness of 25 mm. To obtain the same hull weight, ferrocement would have to be only 8 mm thick. Although small ferrocement boats have been built to this thickness, the impact resistance is not satisfactory for work boats used in fishing or transport. At its present stage of development, ferrocement has proved most suitable for boats above 10 m. Even for this larger size, a ferrocement boat will be heavier than wooden boat, but this is of little disadvantage at moderate speed (between 6.5 and 10 knots).

The boats built in China appear to combine all the favourable characteristics of ferrocement construction. These boats have a length of 15 m with a 10-ton cargo capacity. They are man-powered and thus are operated at a moderate speed. The boats are built on a large scale in a factory using mass-production techniques such as use of precast ferrocement bulkheads and frames that shape the boat and become an integral part of the final structure[10]. These ferrocement sampans are designed for smooth-water use (on rivers and canals) and thus do not require as stringent a design specification as deep-water vessels.

1.5.2 *Service experience*

Ferrocement boats have been in service in large numbers only in the last few years. Many of these boats were not built according to any well-defined specifications. Very little systematic effort has been made to record failures and maintenance problems. Thus, it is difficult to evaluate the performance of ferrocement boats which have been recently built. Nonetheless, some attempts have been made to assess the conditions of existing ferrocement craft[82–84]. Ferrocement fishing boats built in New Zealand have been surveyed by the New Zealand Marine Department and some results of that survey are reported in reference 82. Experience of 300 commercially built crafts in more than 20 countries has been analysed in reference 83. Sutherland, in reference

84, describes his personal experience with about 80 boats which have been in service for up to about 10 years. From these reports, there is clear evidence that ferrocement can be utilized very satisfactorily for boat building. However, the following problem areas are worth noting.

(1) Mortar application and penetration seems to be the greatest problem in ferrocement boat construction. Many defects and failures can be attributed to improper plastering. Several instances in which damage to the hull had revealed serious corrosion of steel were due to the presence of voids; no corrosion of steel was found in dense mortar.
(2) The lack of resistance to impact and punching appear to be the most common cause of damage to ferrocement vessels. Although the repair of the damaged area is not complicated, the frequency and the nuisance of repairing the damage have been reported to reduce the enthusiasm for ferrocement boats among many fishermen[82].
(3) Paint failures seem to be frequent and no single paint formula appears to give consistently satisfactory results[84].

1.5.3 *Silos*

Inadequate storage facilities for grain exist in farms and villages in most developing countries. It has been reported that up to 25% of rice is lost to birds, fungi, rodents and insects in Thailand[85]. Ferrocement silos to store up to 30 tons of grain appear quite suitable and economical for the developing countries. Ferrocement is watertight and, with appropriate sealants, can be made airtight. In an airtight ferrocement bin micro-organisms present cannot survive to damage the stored product[85,92,93].

The Applied Scientific Research Corporation of Thailand[81] has developed a family of economical airtight ferrocement bins which can hold up to 10 tons of grain, other foodstuff, fertilizer, cement, pesticide or up to 5000 gallons of drinking water. The base of the silo is saucer-shaped and consists of two layers of 5 cm-thick concrete with mesh reinforcement and an asphalt seal between. These silos are conical in shape. The curved walls slope inwards to a central entrance hatch at the top. The need for a roof structure is eliminated.

In certain parts of Ethiopia, underground pits are the traditional method of grain storage. It has been found that when the traditional pit is lined with ferrocement and provided with an improved airtight lid, a hermetic and waterproof storage chamber can be achieved[86].

Small-capacity ferrocement bins (up to 3 tons) have been developed, analysed and tested in India[87,92,93]. The units are cylindrical in shape (120 cm in diameter) and are prefabricated in heights of 100 cm. The bins were tested in a laboratory with wheat to ascertain the efficiency of ferrocement in carrying the bin wall load. They were also tested in the field to judge their effectiveness against insects. A cost estimate of ferrocement bins showed them to be cheaper than steel, reinforced concrete or aluminium bins.

1.5.4 *Tanks*

Remarks made above regarding the need in developing countries for efficient grain storage also apply to the storage of drinking water. Thus ferrocement might also be considered for water-tanks in developing countries.

Ferrocement appears also to be an attractive alternative material for the storage of water in the industrially developed countries. Small ferrocement tanks of less than 5000 gallons capacity (20 000 litres) are being factory-built in New Zealand[88]. A recent feasibility study found ferrocement tanks to be less expensive than steel or fibreglass tanks[80]. For the storage of water heated by solar energy, tanks of approximate total capacity of 160 000 gallons (600 000 litres) were required for a proposed design of a new physical science education centre for the Science Museum of Virginia in Richmond, Virginia. A feasibility study was conducted to determine whether there were any advantages to using ferrocement rather than steel or fibreglass for the construction of these tanks. The results of this feasibility study, including the detailed design, methods of construction and drawings for the ferrocement tanks are given in the final report submitted by the Science Museum of Virginia to the U.S. Energy, Research and Development Administration[80]. The following conclusions were drawn based on the study.

(1) Ferrocement is an economically feasible material for the construction of water-storage tanks for storing energy collected by a solar system.
(2) Flexibility of shape, freedom from corrosion, possibility of hot storage, relative lack of maintenance and ductile mode of failure are important advantages of ferrocement over other materials commonly used for low to medium pressure (up to 50 psi $\simeq$ 3.52 kg/cm^2) storage of fluids.
(3) Ferrocement tanks require less energy to produce than those made of steels.

1.5.5 *Roofs*

Of the desperately needed new materials and construction methods in the developing countries, the most critical component is appropriate roofing. Walls and floors of a dwelling can generally be built out of local materials. It has not, however, been possible to manufacture from local materials roofs which are durable, long-lasting and resistant to fire, insects, flood or earthquake. As a result many developing countries import galvanized iron sheets or use asbestos cement sheets; such roofing materials may cost as much as 60% of the total cost of a house[10]. The previously described advantages of ferrocement for boats apply equally well to roofs. Ferrocement roofing material can be factory mass-produced in prefabricated form, a process best suited to the concentrated demands of the urban area. Or it can also be fabricated *in situ* in villages[92,93].

The success of ferrocement for self-help housing is well documented in reference 89 where construction of hundreds of ferrocement roofs for poorer areas of Mexico are described. Many of these ferrocement roofs were dome-shaped with a span of 3–6 m. The domes were shaped by reinforcing rods which were tied to the masonry walls. The construction of the roofs required no mechanical equipment.

There has recently been construction of larger ferrocement roofs in Italy. A recent design and construction of six ferrocement shells for a roof to shelter animals is described in reference 90. The roofs have a span of 17 m and thickness of 3 cm. Note that since the stresses produced by dead load are critical in the design of roofs, thin ferrocement shells should be economical for roof structures.

Corrugated ferrocement sheets have been developed and tested by the Building Research Institute, State Engineering Corporation, Colombo, Sri Lanka. These sheets are developed as a replacement for asbestos cement corrugated sheets which are widely used as a roofing material. It has been reported that ferrocement sheets are less expensive and require less capital investment and foreign exchange. These ferrocement sheets are so designed that their weight, dimensions and load-carrying capacities are similar to the widely-used asbestos cement sheets. In addition, it has been observed that ferrocement sheets are more ductile than asbestos cement sheets[91]. Thus, since the supply of asbestos fibres is limited and since they may be carcinogenic, ferrocement may be a suitable replacement.

1.6 Conclusions

The widespread use of ferrocement has begun only in the last two decades. Thus the state-of-the-art of ferrocement is still in its infancy. Nevertheless, sufficient design information is available and adequate field experience has been acquired to enable safe design and construction of many types of ferrocement structures. Whether ferrocement can economically compete with other alternative materials depends on its type and the country of application. For industrially developing countries where the cost of materials is relatively higher than the cost of labour, some applications of ferrocement appear especially attractive. They include boats, silos, tanks and roofs. For industrially developed countries, ferrocement seems economical for medium size storage tanks, some types of roof shell construction and wherever the ease of forming complicated shapes and the lighter weight of ferrocement can be safely exploited.

Notation

A	Value of Y at maximum load
A_c	Area of the composite
A_{si}	The total area of the ith layer of reinforcement

a_c	Crack spacing
B	A constant
b	Width of the beam
C_m	Total compressive force contribution of mortar
C_s	Total compressive force contribution of reinforcement
c	Depth of neutral axis of the cracked section
d	Diameter of the mortar cylinder specimen
d_i	Distance of the ith layer reinforcement from the extreme compression fibre.
$d_{\max}$	Distance of the extreme tension layer reinforcement from the extreme compression fibre.
E_c, E_m, E_R	Moduli of elasticity for the composite, mortar and reinforcement respectively
e	Base of Napierian logarithms
f'_c	Compressive strength of the mortar
f_m	Maximum stress in mortar
f_r	Modulus of rupture
f_s	Reinforcement stress
f_{si}	Stress in the ith layer of reinforcement
$f_{s\max}$	Stress in the outermost layer of steel
f_{sr}	Stress range in the outermost layer of steel
f_{tm}	Split tensile strength of mortar
f_y	Yield strength of steel
h, t	Thickness of the specimen
I	Moment of inertia
I_g, I_{cr}, I_e	Moments of inertia of the gross, cracked and effective sections respectively
I_{ec}	Effective moment of inertia of the cracked region of the beam
jd_i	Lever arm of the ith layer of reinforcement measured from the neutral axis.
k	A constant
L	Span of the beam
l	Length of the specimen
M	Moment
M_a	Maximum moment along the span
M_{cr}	Cracking moment
m	Number of layers of wire meshes
N	Number of cycles
N_f	Number of cycles to failure
n	Modular ratio, E_R/E_m
P	Load
r	N/N_f or a constant
S_R	Lateral surface of the reinforcement per unit volume of the composite

S_{RL}	S_R in the loading direction
S	Mesh size
T_m	Total tensile force contribution of mortar
T_s	Total tensile force contribution of reinforcement
t	Same as h
t'	Distance in which meshes are distributed
V_R, V_{RL}	Volume fraction of the reinforcement, subscript L refers to the longitudinal direction
W_{av}	Average crack width
W_{max}	Maximum crack width
Y	Deflection, average or maximum crack width
β	Ratio of distances to the neutral axis from the extreme tensile fibre and from outermost layer of steel
ε_c	Strain in the extreme compression fibre
ε_s	Strain in the extreme tension layer of reinforcement
ε_{si}	Strain in the ith layer of reinforcement
ε_t	Strain in the extreme tension layer
$(\Delta l)_{av}$	Average crack spacing
δ	Deflection
ϕ	Diameter of the wire
θ	Average crack spacing/maximum crack spacing
η	Bond strength/tensile strength
ψ	Curvature
ψ_y	Curvature when the extreme tension layer of reinforcement starts yielding.

References

1. Biggs, G.W. (1975) Bamboo reinforced ferrocement grain storage silo. *J. Struct. Eng.* (India) **2**, 173–182.
2. Sharma, P.C., Pama, R.P., Valls, J., and Robles-Austriaco (1981) 'Ferrocement applications for rural development in Asian Pacific countries', *Proc. Int. Symp. on Ferrocement*, (eds. Shah, S.P and Oberti, G.) RILEM, 3.117–3.126.
3. Ellen, P.E. (1972) 'Practical ferrocement design, reinforced and post-tensioned', *FAO Seminar on the Design and Construction of Ferrocement Fishing Vessel*, Wellington, New Zealand, 26.
4. Alexander, D.J. (1975) Techniques in the construction of prestressed concrete barges and prestressed ferrocement vessels. *J. Ferrocement* **4**, 2–10 and 13–15.
5. Nervi, P.L. (1956) *Ferro-Cement Structures*, F.W. Dodge Corporation, New York, 50–62.
6. Nervi, P.L. (1953) Precast concrete offers new possibilities for design of shell structures. *J. Amer. Conc. Inst.* 50.
7. Nervi, P.L. (1951) Ferrocement: its characteristics and potentialities. *L'Ingegrnere*, Italy, English translation by Cement and Concrete Association, No. 60, London, 1956, 17.
8. Cassie, W.F. (1967) Lambot's boats—a personal discovery. *Concrete* (London) **1**, 380–382.
9. Morgan, R.G. (1968) Lambot's boats. *Concrete* (London) **2**, 128.
10. *Ferrocement: Applications in Developing Countries* (1973) National Academy of Sciences, 90.
11. *Ferrocement—Materials and Applications* (1979) Publication SP-61, American Concrete Institute, 195.
12. ACI Committee 549 (1982) State-of-The-Art Report on Ferrocement, *Concrete International* **4**, 13–38.

13. Paul, B.K., and Pama, R.P. (1978) *Ferrocement*. International Ferrocement Information Centre Publication, 149.
14. *Proc. Int. Conf. on Materials of Construction for Developing Countries*, 1978, International Ferrocement Information Centre Publication.
15. Sharma, P.C., and Gopalaratnam, V.S., *Ferrocement Grain Storage Bin, Ferrocement Water Tank, Ferrocement Biogas Holder* and *Ferrocement Canoe*, International Ferrocement Information Centre Publications.
16. Shah, S.P., and Oberti, G. (eds.) (1981) *Proc. Int. Symp. on Ferrocement* 440, available from ISMES, Viale Giullo Cesare 29, 24100 Bergamo, Italy.
17. Shah, S.P. (1983) Recent developments in ferrocement. *RILEM J. Mater. Struct.*, 341–347.
18. Shah, S.P. (1974) New reinforcing materials in concrete construction. *J. Amer. Concr. Inst.* **71**, 257–262.
19. Naaman, A.E., and Shah, S.P. (1971) Tensile tests on ferrocement. *J. Amer. Concr. Inst.* **68**, 693–698.
20. Johnston, C.D., and Mattar, S.G. (1976) Ferrocement behavior in tension and compression. *J. Struct. Div. (ASCE)* **102**, 875–899.
21. Somayaji, S., and Shah, S.P. (1981) 'Prediction of tensile response of ferrocement', *Proc. Int. Symp. on Ferrocement*; eds. Shah, S.P. and Oberti, G., RILEM, 1.73–1.84.
22. Somayaji, S., and Shah, S.P. (1981) Bond stress-slip relationship and cracking reference of tension member. *J. Amer. Concr. Inst.* **78**, 217–225.
23. Raju, N.K. (1970) Comparative study of the fatigue behavior of concrete, mortar, and paste in uniaxial compression. *J. Amer. Concr. Inst.* **67**, 461–463.
24. Saito, M., and Imai, S. (1983) Direct tensile fatigue of concrete by the use of friction grips. *J. Amer. Concr. Inst.* **80**, 413–438.
25. Shah, S.P. (1970) *Ferrocement as a New Engineering Material*, Publication for the Annual Convention of the Canadian Section of ACI, Department of Materials Engineering, University of Illinois at Chicago Circle, 37.
26. Logan, D., and Shah, S.P. (1973) Moment capacity and cracking behavior of ferrocement in flexure. *J. Amer. Concr. Inst.* **70**, 799–804.
27. Suryakumar, G.V., and Sharma, P.C. (1975) An investigation into the flexural behavior of ferrocement. *J. Struct. Eng.* (India) **2**, 137–144.
28. Rajasekaran, S., Raju, G., and Palanichamy, K. (1975) Behavior of ferrocement specimens in bending and compression. *J. Struct. Eng.* (India) **2**, 145–154.
29. Rajagopalan, K., and Parameswaran, V.S. (1975) Analysis of ferrocement beams. *J. Struct. Eng.* (India) **2**, 155–164.
30. Johnston, C.D., and Mowat, D.N. (1974) Ferrocement-material behavior in flexure. *J. Struct. Div. (ASCE)* **100**, 2053–2069.
31. Greenius, A.W. (1975) *Ferrocement for Canadian Fishing Vessels*, Technical Report, Industrial Development Branch, Publication No. 86, Vancouver, Canada, 160.
32. Austriaco, N.C., Lee, S.L., and Pama, R.P. (1975) Inelastic behavior of ferrocement slabs in bending. *Mag. Concr. Res.* **27**, 193–209.
33. Balaguru, P.N. Naaman, A.E., and Shah, S.P. (1977) Analysis and behavior of ferrocement in flexure. *J. Struct. Div. (ASCE)* **103**, 1937–1951.
34. Guerra, A.J., Naaman, A.E., and Shah, S.P. (1978) Ferrocement cylindrical tanks: cracking and leakage behavior. *J. Amer. Concr. Inst.* **75**, 22–30.
35. Shah, S.P., and Naaman, A.E. (1978) Crack control in ferrocement and its comparison with reinforced concrete. *J. Ferrocement* **8**, 67–80.
36. Naaman, A.E. (1979) *Design Predictions of Crack Widths in Ferrocement*, Special Publication 61, American Concrete Institute, 25–42.
37. Balaguru, P.N. (1981) Predicting crack widths in ferrocement beams. *J. Ferrocement* **11**, 203–214.
38. Picard, A., and Lachance, L. (1974) Preliminary fatigue tests on ferrocement plates. *Cem. Concr. Res.* **4**, 967–978.
39. Balaguru, P.N., Naaman, A.E., and Shah, S.P. (1979) Fatigue behavior and design of ferrocement beams. *J. Struct. Div. (ASCE)* **105**, 1333–1346.
40. Balaguru, P.N., Naaman, A.E., and Shah, S.P. (1979) Serviceability of ferrocement subjected to flexural fatigue. *Int. J. Cem. Composites* **1**, 3–9.
41. Singh, G. (1981) 'Flexural response of a ferrocement to constant deflection repeated loading', *Proc. Int. Symp. on Ferrocement*, RILEM, 2.1–2.6.

42. Balaguru, P.N. (1981) Cracking behavior of ferrocement beams under static and fatigue loading. *Proc. Int. Symp. on Ferrocement*, (eds. Shah, S.P., and Oberti, G.) RILEM, 2.127–2.137.
43. Shah, S.P., and Key, W.H., Jr. (1972) Impact resistance of ferrocement. *J. Struct. Div. (ASCE)* **98**, 111–123.
44. Srinivasa Rao, P.S., Achyatha, M.S., Mathews, M.C., and Srinivasan, P.P. (1981) 'Impact studies on ferrocement slabs', *Proc. Int. Symp. on Ferrocement* (eds. Shah, S.P., and Oberti, G.), 1.7–1.19.
45. Swamy, R.N., and Al-Wash, A.A. (1981) 'Cracking behaviour of ferrocement in flexure'. *Proc Int. Symp. on Ferrocement* (eds. Shah, S.P., and Oberti, G.) A/1–A/11.
46. Irons, M.E., and Watson, L.L. (1977) Ferrocement boats reinforced with expanded metal. *J. Ferrocement* **7**, 9–16.
47. ACI Committee 318, 1983, *Building Code Requirements for Reinforced Concrete*, American Concrete Institute, **103**, and *Commetary on Building Code Requirements for Reinforced Concrete*, Americam Concrete Institute, 132.
48. Branson, P.E. (1965) Design procedures for computing deflections. *J. Amer. Concr. Inst.* **63**, 730–740.
49. New Zealand Marine Department (1975) Requirements for the construction of ferrocement boats. *J. Ferrocement* **4**, 14–16.
50. Eyers, D.J. (1975) Review of 'Det Norske Veritas: requirement for ferrocement boats', *J. Ferrocement* **4**, 26.
51. Kowalski, T.K. (1975) Draft mortar specifications. *J. Ferrocement* **4**, 29–30.
52. Bangh, I., Bowen, G.L., and Kenyon, G. (1975) The ferrocement handbooks: an introduction to the ferrocement yacht. *J. Ferrocement* **4**, 28.
53. Samson, J., and Geoff, W. (1968) *How to Build a Ferrocement Boat*, Samson Marine Enterprises, Ltd., Canada.
54. Shaw, H.F. (1970) *Ferrocement on Structural Materials at Cryogenic Temperatures*, M.S. thesis, Massachusetts Institute of Technology.
55. Scott, W.G. (ed.) (1971) *Ferrocement for Canadian Fishing Vessels.* Industrial Development Branch, Fisheries Service, Department of Environment, Ottawa, Canada.
56. Biggs, G.W. (1972) *An Introduction to Design for Ferrocement Vessels.* Industrial Development Branch, Fishing Division, Department of Environment, Ottawa, Canada.
57. Bezukladov, V.F., Amel'Yanovich, K.K., Verbitskiy, V.D., and Bogoyavlenskiy, L.P. (1968) Ship hulls made of reinforced concrete. *NAVSHIPS Translation* No. 1148.
58. Walkus, I.R., and Kowalski, T.G. (1971) Ferrocement: A survey. *Concrete* (*London*) 48–58.
59. Shah, S.P., and Srinivasan, M.G. (1973) *Strength and Cracking of Ferrocement.* Fishing News Ltd., West Byfleet, Surrey, England.
60. American Society for Testing and Materials (1980) *1980 Annual Book of ASTM Standards, Part 14*, Concrete and Mineral Aggregates (including Manual of Aggregate and Concrete Testing), American Society for Testing and Materials, 1916 Race St., Philadelphia, PA 19103, U.S.A., 834.
61. Lea, F.M. (1956) *The Chemistry of Cement and Concrete.* Revised edn. of Lea and Desch, St. Martin's Press, New York, 637.
62. Troxell, G.E., Davis, H.E., and Kelly, J.W. (1968) *Composition and Properties of Concrete.* 2nd Edn. McGraw–Hill Book Co., New York, 529.
63. Powers, T.C. (1968) *The Properties of Fresh Concrete.* John Wiley, New York, 59 and 299.
64. Popvics, S. (1979) *Concrete-Making Materials.* McGraw-Hill Book Co., New York, 370.
65. Irons, M.E. (1980) Some improved methods for building ferrocement boats. *J. Ferrocement* **10**, 189–203.
66. Irons, M.E. (1982) Tips for amateur builders. *J. Ferrocement* **12**, 229–293.
67. Shah, S.P. (1979) Tentative recommendations for the construction of ferrocement tanks. *Amer. Concr. Inst.*, SP-61, 103–113.
68. Raichvarger, Z., and Raphael, M. (1979) Grading design of sand for ferrocement mixes, *Amer. Concr. Inst., SP-61*, 115–131.
69. Ramakrishnan, V. (1977) Significant physical and mechanical properties of gap graded concrete. *Indian Concr. J.* **51**, 142–148.
70. Dinsenbacher, A.L., and Brauer, F.E. (1974) Material development, design, construction and evaluation of a ferrocement planning boat. *Marine Technol.* **11**, 277–296.

71. Collen, L.D.G. (1960) Some experiments in design and construction with ferro-cement. *Trans. Inst. Civil Eng. Ireland* **86**, 40.
72. Byrne, J.G., and Right, W. (1961) An investigation of ferro-cement using expanded metal. *Construction Eng.* 429.
73. Atcheson, M., and Alexander, D. (1979) Development of fibrous ferrocement. *Amer. Concr. Inst.*, SP-61, 81–101.
74. ACI Committee 544 (1982) State-of-the-art report on fiber reinforced concrete. *Concrete International: Design and Construction* **4**, 9–30.
75. Shah, S.P. (1983) 'Fiber reinforced concrete', *Handbook of Structural Concrete*, (eds. Kong, Evans, Cohen and Roll), McGraw-Hill, chap. 6.
76. Balaguru, P.N., Naaman, A.E., and Shah, S.P. (1976) *Ferrocement in Bending; Part I: Static Nonlinear Analysis*, Report No. 76–2, Department of Materials Engineering, University of Illinois at Chicago Circle.
77. American Society for Testing and Materials (1980) *1980 Annual Book of ASTM Standards, Part 4*, Structural Steel: Concrete Reinforcing Steel; Pressure Vessel Plate and Forgings; Steel Rails, Wheels and Tires; Fasteners, American Society for Testing and Materials, 1916 Race St., Philadelphia, Pa.
78. Martin, E.I., and Watson, L.L., Jr. (1972) U.S. Patent 3,652,755, issued March 28, 1972.
79. Study and Evaluation of Ferrocement for Use in Wind Tunnel Construction, Report JABE-ARC-07, Research Contract No. NA52-5889, NASA Ames Research Center, Moffett Field, California, July 1972, 88.
80. *An Integrated Solar Energy Heating and Cooling System for a New Physical Science Education Center in Richmond Virginia*, Final report prepared for U.S. Energy Research and Development Administration, Division of Solar Energy, Contract No. E-(40–1), 4899.
81. Pama, R.P., Lee, S.L., and Vietmeyer, N.D., (eds.) (1974) *Ferrocement, A Versatile Construction Material: Its Increasing Use in Asia*, Report, Workshop on the Introduction of Technologies in Asia, Asian Institute of Technology, Bangkok, 106.
82. Eyers, J.D. (1973) *Survey of Ferrocement Fishing Boats Built in New Zealand, FAO Investigates Ferrocement Fishing Craft*. Fishing News Ltd., West Byfleet, Surrey, England.
83. Hagenbach, T.M., *Experience of* 300 *Commercially Built Craft in More Than* 200 *Countries. ibid.*
84. Sutherland, W.M. *Ferrocement Boats—Service Experience in New Zealand, ibid.*
85. Smith, R.B.L., and Boon-Long, S. (1971) Hermetic storage of rice for Thai farmers. *Thai J. Agr. Sci.* **4**, 143–155.
86. Hall, D.W. (1971) *Handling and Storage of Food Grains in Tropical and Subtropical Countries.* Food and Agricultural Organization of the United Nations, Rome, 181.
87. Structural Engineering Research Centre (1975) *A Proposal for Production of Small Capacity Ferrocement Bins (*1, 2 *and* 3 *tons)*. Report No. 5/C-15, Roorkee (U.P.), India.
88. *Ferrocement Tanks and Utility Buildings* (1968) Bulletin No. CP-10, New Zealand Portland Cement Association, Wellington.
89. Castro, J. (1977) Ferrocement roofing manufactured on a self-help basis *J. Ferrocement* **7**, 17–27.
90. Barberio, V. (1975) Cupulas Delgadas De Ferrocemento Para Una Instalacion Ictica En El Rio Tirino. *J. Mexican Concr. Inst.*, Revista IMCYC, Vol. XIII, No. 74, 20–28.
91. Naaman, A.E., and Shah, S.P. (1976) 'Evaluation of ferrocement in some structural applications', *Proc. Int. Symp. on Housing Problems*, Atlanta, U.S.A., 1069–1085, Florida International University, International Institute for Housing and Building, Tamiami Trail, Miami, Florida.
92. Subrahmanyam, B.V., and Abdul Karim, E. (1979) Ferrocement technology: a critical evaluation. *Int. J. Cem. Composites* **1**, 125–140.
93. Pama, R.P., and Gopalaratnam, V.S. (1979) Ferrocement—applications, research and developments. *Int. J. Cem. Composites* **1**, 159–169.
94. *Concrete International*, Design and Construction (1983) American Concrete Institute, 5, 9–50.

2 Fibre Reinforced Concrete

J.G. KEER

Synopsis

The inclusion of fibre reinforcement in concrete, mortar and cement paste can enhance many of the engineering properties of the basic materials, such as fracture toughness, flexural strength and resistance to fatigue, impact, thermal shock or spalling. Fibre reinforcement is likely to be used (in preference to conventional reinforced or prestressed concrete) if these properties can be exploited in conjunction with advantages in construction or fabrication techniques, e.g. the inclusion of reinforcement as an integral part of the fresh concrete, or the manufacture of thin sheet products.

In this chapter, the basic concepts of fibre reinforcement of brittle matrices are reviewed. Important effects of fibres of, principally, steel, glass and polypropylene upon the properties of concrete, mortar and cement paste are discussed.

The fibre reinforced concrete 'industry' has developed along two basically separate paths, i.e. bulk fibre-reinforced concrete and thin sheet products of fibre-reinforced cement or mortar. Applications in both areas are given and possible future developments are outlined.

2.1 General background

2.1.1 *Introduction*

Cementitious materials in the form of mortars or concretes are attractive for use as constructional materials since they are cheap, durable and have adequate compressive strength and stiffness for structural use. Additionally, in the fresh state they are readily moulded so that complex shapes may be fabricated. Their deficiencies lie in their brittle characteristics—poor tensile and impact strength—and in their susceptibility to moisture movements. Reinforcement by fibres can offer a convenient, practical and economical method of overcoming these deficiencies, particularly in applications where conventional reinforcement by steel bars, carefully positioned to obtain maximum benefit from the reinforcement, is unsuitable. Furthermore, the provision of small-size reinforcement as an integral part of the fresh concrete

Table 2.1 Fibre and matrix properties

Fibre/matrix	*Density (kg/m³)*	*Young's modulus (GN/m²)*	*Tensile strength (MN/m²)*	*Elongation at break (%)*
Steel	7860	200	1000–3000	3.5
Glass (Cem-FIL filament)	2700	80	2500	3.5
Polypropylene	910	3–15	400	8
Cellulose	1500	10–40	500	—
Carbon (high modulus)	1900	380	1800	~0.5
Carbon (high strength)	1900	230	2600	~1.0
Kevlar (PRD 49)	1450	133	2900	2.1
High modulus polyethylene	960	15–40	400	3
Asbestos (chrysotile)	2550	164	200–1800 (fibre bundles)	2–3
OPC paste	2000–2200	10–30	3–8	0.01–0.05
OPC concrete	2300	30–40	1–4	0.005–0.015

mass can provide advantages in terms of the fabrication of products and components.

To date, excluding the use of asbestos, the fibres finding greatest use have been steel and polypropylene, principally in concrete, and glass, principally in cement mortar for thin section applications. In this last field, considerable development has also taken place in the use of polypropylene as reinforcement and the chapter, therefore, concentrates on these three man-made fibres—steel, glass, and polypropylene. The use of natural cellulose fibres is also considered because of their importance as direct replacements for asbestos in asbestos-cement products.

The reinforced matrix is generally based on ordinary portland cement (OPC), but alternative or less straightforward matrices may be used. Typical property values for cementitious matrices and a number of reinforcing fibres are given in Table 2.1.

2.1.2 *Development of fibre concrete technology*

The modern technology of the reinforcement of brittle cement and concrete matrices by fibres, which has developed rapidly over the last twenty years, was initiated with the invention of asbestos-cement by Ludwig Hatschek in 1899. Hatschek had experimented with the reinforcement of asphalt and oxychloride cements by asbestos fibres before he adopted the portland cement binder that was to prove such a huge commercial success[1]. The principal activities associated with fibre-cement technology during the early years of the

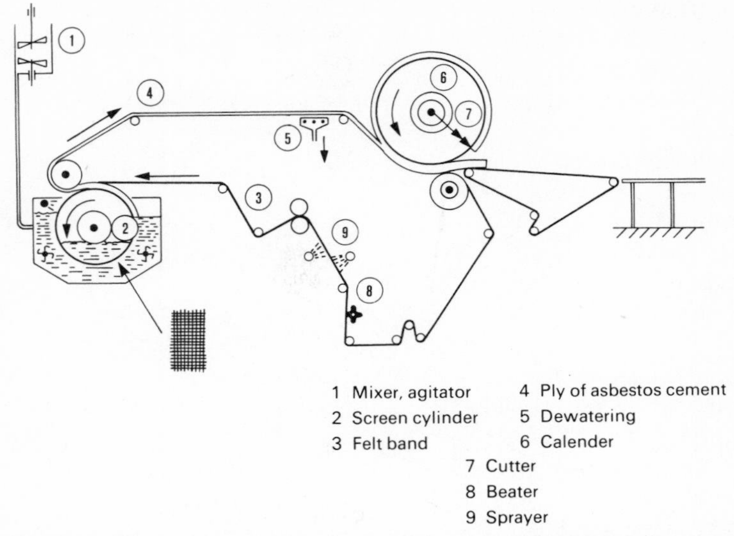

Figure 2.1 The manufacture of asbestos-cement sheets by the Hatschek process

twentieth century appear to have involved variations in process or composition in order to circumvent Hatschek's patent. Alternative fibre cements or production processes, however, remained comparatively unimportant.

In the Hatschek, or wet, process, an asbestos-cement sheet is built up to the desired thickness by a lamination process involving successive applications of a thin wet layer of asbestos and cement particles. The sheet is moulded while still pliable into the required shape. A schematic representation of the manufacture of flat and corrugated asbestos-cement sheet is given in Fig 2.1. It is important to note that the asbestos fibres have a further key function, in addition to reinforcing, in controlling slurry drainage during manufacture to ensure that the water content of the slurry is reduced without segregation of the cement. Any alternative fibre considered by asbestos-cement manufacturers must serve the same function if Hatschek process machinery (a considerable capital investment) is to continue in use. Indeed the absorptive capacity of asbestos for cement was formerly central to a 1930s definition of a fibre cement[2]: 'Fibre cement is an artificial stone made of fibres and binder and, if required, cement fine additions, in the preparation of which sufficient fibres are used to ensure their uniform distribution in the artificial stone as a result of their absorptive capacity for cement fine materials.'

Asbestos-cement became extensively used as a material for roofing and cladding, in flat and corrugated sheet form, and for pipe manufacture. In 1974 for example, one source reported that approximately 8.5. million tonnes of asbestos cement were produced outside of the USSR, Eastern Bloc countries and China[3]. Data from these countries are unreliable but production is probably of the same order.

Despite a number of patents for glass and steel-fibre reinforced cement and concrete in the 1920s and 1930s, asbestos-cement remained the dominant fibre-cement/concrete technology until the 1960s.

In 1963, however, Romualdi and Batson[4] published the results of an investigation carried out in the U.S.A. on steel-fibre reinforced concretes, which stimulated considerable activity in fibre concrete research and resulted in a number of applications of steel-fibre concrete, e.g. in pavement construction, in the early 1970s. There followed in 1964 the publication of pioneering work on the reinforcement of cement and concrete by glass fibres by Krenchel in Denmark[5] and by Biryukovich *et al.*[6] in the USSR. In the latter work, the problem of the attack by alkalis on the 'E' glass fibres had been overcome by the use of alumina cement of low alkaline content. The development of a glass fibre with a sufficient degree of alkali resistance such that it could be used in a portland cement environment was achieved by Dr A.J. Majumdar at the Building Research Establishment in the U.K. in 1967. In collaboration with Pilkington Bros Ltd., the alkali-resistant Cem-FIL fibre was launched which led to world-wide commercial exploitation of glass-reinforced cement (grc).

Polymer fibres such as nylon, polypropylene and polyethylene were investigated in the early 1960s as reinforcement for concrete subject to explosive loading[7]. In 1966, the Shell Company developed a process for producing a fibre concrete, designated Caricrete, containing polypropylene fibres in fibrillated film form, whereby the products, such as driven pile segments, benefited from the improved impact resistance of the concrete. More recently it was realized that with their low price, high strength and ready availability, polymers such as polypropylene also have the potential to increase the tensile strength and strain to failure of a cement-based matrix in competition with, particularly, glass and asbestos fibres in thin-sheet applications.

It is ironic that the extensive efforts of the last twenty years in research and development to produce new fibre cements in thin sections could reap their richest reward, not in advancing the materials into new areas of use, but in replacing the most successful fibre cement ever, asbestos-cement. Following widespread concern about the dangers to health of asbestos, the asbestos-cement industry appears to be in irreversible decline. This has given a further stimulus to fibre cement and concrete technology in the early 1980s and other reinforcements have emerged, e.g. polyvinyl alcohol (PVA) fibres, or re-emerged, e.g. cellulose fibres.

2.2 Theoretical principles of fibre reinforcement

Since the objective of the addition of fibres to cement or concrete is, generally, to improve the mechanical properties of the matrix, it is worth examining the theoretical approaches which have led to an understanding of the manner in which improved properties are achieved.

The mechanics of fibre-cement composites has been discussed extensively in the literature (see for example references 8 and 9) but the following review is based on the method adopted and on the material covered in detail in reference 10.

2.2.1 *The modulus of elasticity of the uncracked composite*

It is usual to assume that the modulus of elasticity, E_c, of a fibre-reinforced cement or concrete prior to matrix cracking is given by the Law of Mixtures, i.e.

$$E_c = \eta_1 \eta_2 E_f V_f + E_m V_m \tag{2.1}$$

where E is the modulus of elasticity, V the volume fraction and subscripts m and f refer to matrix and fibre respectively. η_1 and η_2 are efficiency factors, depending on the orientation and length of the fibre respectively. For continuous fibres aligned in the direction of applied stress $\eta_1 = \eta_2 = 1$. For a random orientation of fibres in two dimensions η_1 is about $\frac{1}{3}$; in three dimensions η_1 may be as small as $\frac{1}{6}$. For practical composites, the length efficiency factor is likely to be close to unity in the region prior to matrix cracking.

The stiffness of a practical composite is unlikely to be greatly improved by the addition of fibres, since V_f in equation (2.1) is generally small (< 0.1) compared to V_m (~ 0.9).

2.2.2 *The failure strain of the matrix*

It follows from above that *if* the matrix failure strain (i.e. the strain at which cracks propagate unstably across the cross-section of a tensile specimen) is unaffected by the presence of fibres, then the cracking stress of a fibre cement or concrete will not be substantially increased by the presence of fibres.

In fact, there has been a considerable debate, which is continuing, over the effect of fibres upon the matrix failure strain. It has already been mentioned that a major stimulus was given to the development of fibre cements and concretes by the work of Romualdi and Batson[4]. By a fracture mechanics approach, they found that, theoretically, the tensile strength of concrete should be significantly increased by the inclusion of closely spaced fibres. The supporting experimental work was based, however, on flexural testing. Subsequently, a number of workers found either slight or no improvment in cracking strengths in tests in direct tension[11] (Fig 2.2). It has since been pointed out by Kelly[12] that Romualdi's theoretical approach resulted in high bond stresses between fibre and matrix which, in practice, could not be sustained.

From a detailed consideration of the energy requirements for a crack to form in the matrix, Aveston *et al.*[13] have shown that the matrix will fail either

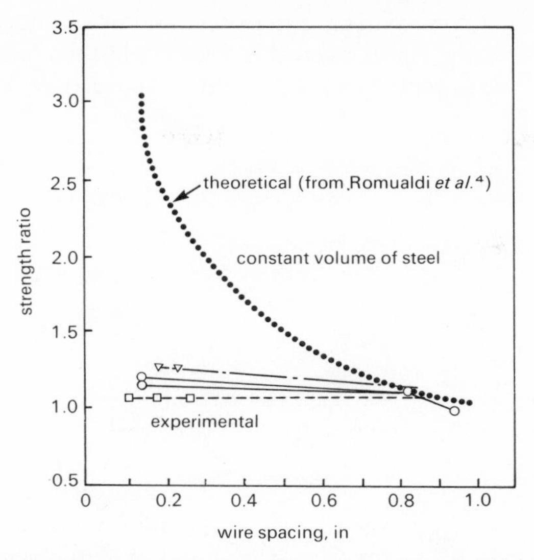

Figure 2.2 Effect of spacing of reinforcement on cracking strength of concrete[11]

when it reaches its normal cracking strain, ε_{mu}, or when the strain reaches a value ε_{muc}, whichever is the greater. ε_{muc} is given by

$$\varepsilon_{muc} = \left\{ \frac{12\tau\gamma_m E_f V_f^2}{E_c E_m^2 r V_m} \right\}^{1/3} \tag{2.2}$$

in which τ is the fibre-matrix frictional stress transfer, γ_m the work of fracture of the matrix and r the fibre radius.

A good correlation between the theory and experimental results for cement paste reinforced with continuous 0.13 mm dia. steel wire or with carbon fibre has been reported[14] (Fig. 2.3). If a normal cracking strain of concrete in direct tension is considered to be about 0.01%, then, according to Fig. 2.3, about 0.5% of continuous steel wires should suffice for crack suppression. For short randomly orientated 3-D wires, the efficiency factor applied to V_f in equation (2.2) will be about 1/6 so that the critical V_f for the beginning of crack suppression for a steel fibre-reinforced concrete may be about 3%. This is on the upper limit of the volume of short fibres that can be included in practice.

More recently, considerations of the mechanics of crack growth and the stabilizing effects of fibres on crack development have led to further theories for the increased failure strain of brittle matrices reinforced by fibres[15,16]. The theory of Aveston *et al.* is now considered to give, in general, a lower limit to the strain that must be exceeded for cracking to occur. This is illustrated in Fig. 2.4 from reference 16 in which the similar crack growth theories of Korczynskyj *et al.* (*K*) and Hannant *et al.* (*H*) and the theory of Aveston *et al.* (*A*) are compared with experimental results of an OPC mortar containing

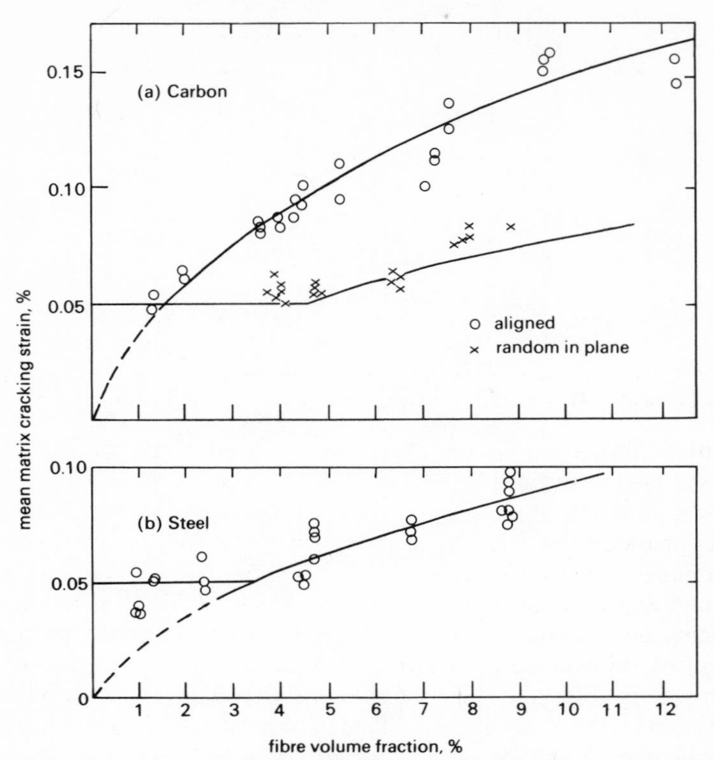

Figure 2.3 Mean matrix cracking strain of steel wire and carbon fibre reinforced cement. Lines are the values computed from equation (2.2)[14]

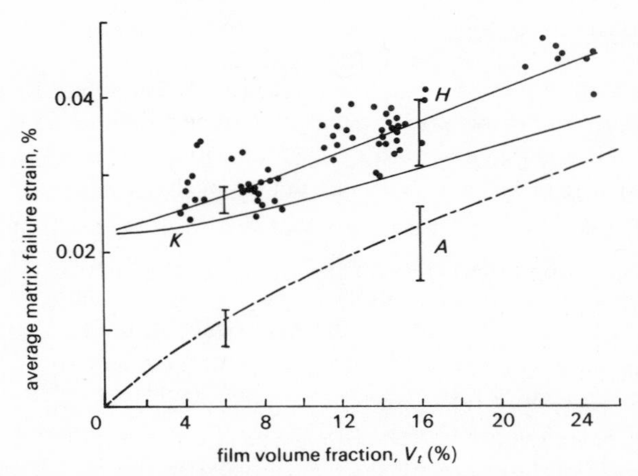

Figure 2.4 Theoretical and experimental values of enhanced matrix failure strain of mortar containing fibrillated polypropylene. Upper and lower bars at $V_f = 6\%$ and 16% indicate the variation of failure strain predicted for values of τ between 0.2 MN/m^2 and 0.8 MN/m^2 (*H*, ref. 16, *K*, ref. 15, *A*, ref. 13)

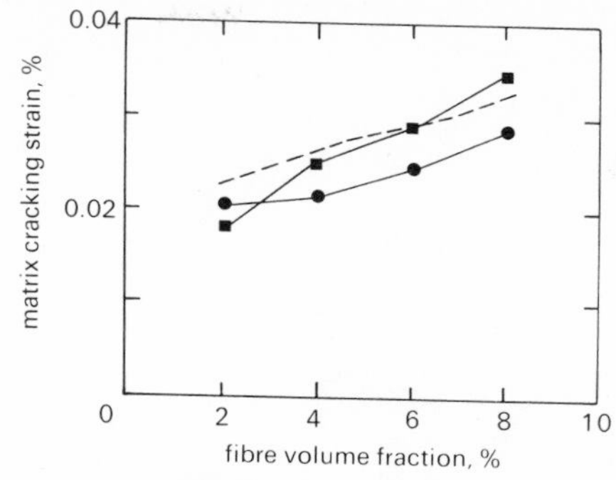

Figure 2.5 Theoretical and experimental values of enhanced matrix failure strain of a glass fibre-reinforced cement[16]. ■, stored in water, ●, stored in air.

continuous, fibrillated polypropylene film. According to the theory of Aveston *et al.*, the failure strain of the unreinforced matrix (0.022%) would not be increased until the fibre volume exceeded about 16%; whereas the other theories predict, more correctly, the enhancement in matrix failure strain with fibre volumes increasing from zero. Figure 2.5 compares theoretical (Hannant *et al.*) and experimental values of enhanced matrix failure strain of cement containing glass fibres randomly orientated in two dimensions. Thus inclusion of about 6% by volume of glass fibres or about 10% by volume of aligned polypropylene fibres may enhance the matrix cracking strain by about 50%. However, experimentally, there is likely to be a considerable scatter of results either side of such an increase, so that care must be taken in using enhanced cracking strains for design purposes.

2.2.3 *Post-cracking behaviour in tension*

The failure strain of a reinforcing fibre is generally considerably greater than that for the matrix and considerable benefit can be gained if full use is made of the ductility of the fibre component. Once the brittle matrix cracks, three types of behaviour in tension may be exhibited by a fibre cement or concrete, as illustrated in Fig 2.6:

(a) The composite fails as fibres fracture under the increased stress thrown onto them (Fig. 2.6(*a*)).
(b) The composite can carry a decreasing load as the fibres pull out from the cracked surfaces (Fig. 2.6(*b*)). No increase, after matrix cracking, of the tensile strength of the composite is observed, yet the strain at complete failure is increased and there can be a considerable increase in the toughness of the composite as measured by the area under the complete stress–strain curve. This type of behaviour is typical of some short, randomly orientated steel or organic fibre composites.
(c) The composite continues to carry an increasing tensile stress. Multiple

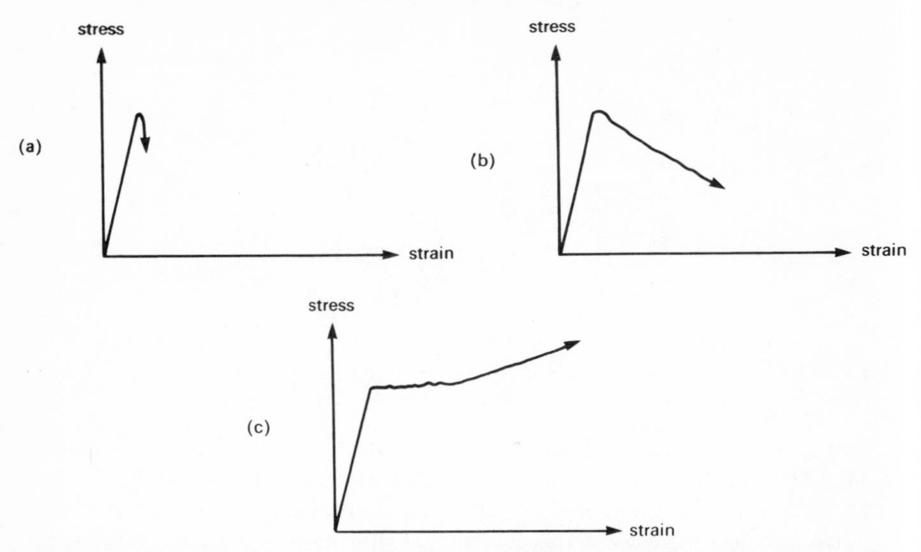

Figure 2.6 Types of behaviour in tension exhibited by fibre cements or concrete: (*a*) composite fails when matrix cracks; (*b*) composite carries a decreasing load as fibres pull out at crack; (*c*) composite can carry an increasing load after matrix cracking

> cracking of the matrix occurs and the material behaves in a pseudo-ductile fashion with a high impact strength (Fig. 2.6(*c*)). Cement or mortar matrices with a sufficient volume of continuous (or long) fibres and glass-fibre reinforced cement may exhibit this type of behaviour.

For multiple cracking to occur, the fibre volume must be greater than a critical volume given by:

$$V_{f\,\mathrm{crit}} = \frac{E_c \varepsilon_{mu}}{\sigma_{fu}} \tag{2.3}$$

where ε_{mu} is the strain at which the matrix cracks and σ_{fu} is the failure stress of the fibres or the stress which causes the fibres to pull out of the matrix. It should be noted that the critical fibre volume may increase substantially with time, since the matrix may increase in modulus and cracking strain and the fibre decrease in strength. Thus a composite that is initially tough and ductile, with a tensile stress–strain characteristic as in Fig. 2.6(*c*), may change to a characteristic such as Fig. 2.6(*a*) with a brittle failure when a single crack forms. Glass-reinforced cement may exhibit this change after ageing in wet storage or after natural weathering.

An idealised form of the tensile stress–strain curve for a composite exhibiting *multiple cracking* has been presented by A veston *et al.*[13] and is shown in Fig. 2.7. It is assumed that the fibres are continuous and aligned, that the bond between fibres and matrix is purely frictional and that the matrix has a well-defined single-valued breaking stress.

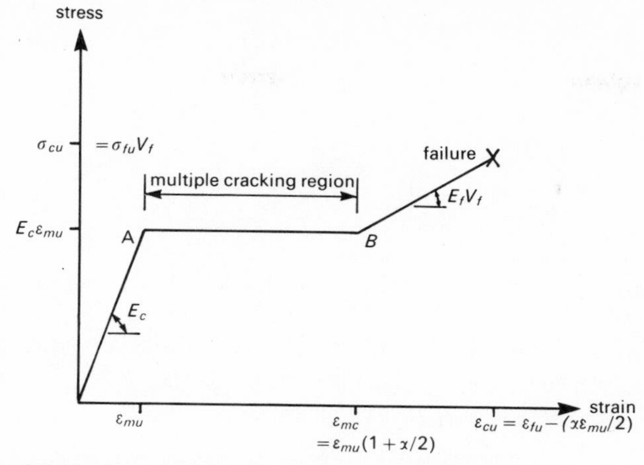

Figure 2.7 Idealized tensile stress–strain curve of a fibre cement or concrete[13]

When the matrix cracks, then provided $V_{f\,\mathrm{crit}}$ is exceeded, the additional load is transferred back into the matrix over a transfer length x' given by:

$$x' = \frac{V_m}{V_f}\frac{\sigma_{mu}}{\tau}\frac{A_f}{P_f} \tag{2.4}$$

where $\sigma_{mu}(=E_m\varepsilon_{mu})$ is the matrix failure stress, A_f is the fibre cross-sectional area and P_f the fibre perimeter.

Eventually the matrix will be broken down into a series of blocks of lengths between x' and $2x'$, as illustrated in the multiply cracked polypropylene-reinforced specimen of Fig. 2.8. For the case when the crack spacing is $2x'$, the additional stress on the fibres when the matrix cracks will vary between $\sigma_{mu}V_m/V_f$ at the crack surface and zero at x' from the crack. The average additional strain in the fibres, $\Delta\varepsilon_c$, is equal to the extension per unit

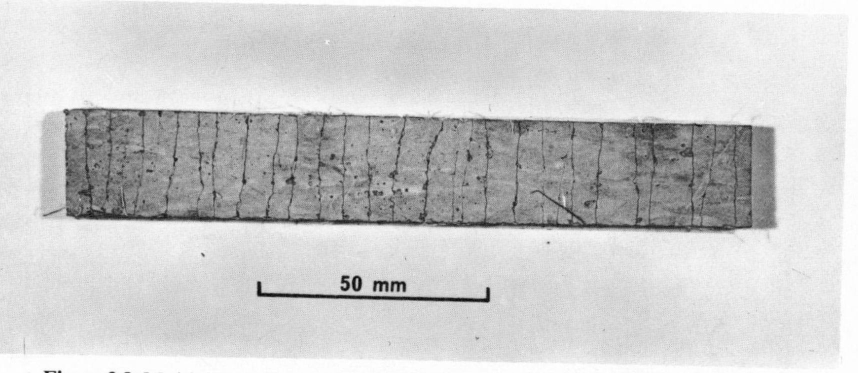

Figure 2.8 Multiple cracking of a polypropylene-reinforced cement composite

length of composite at constant stress, $E_c \cdot \varepsilon_{mu}$, and is given by:

$$\Delta\varepsilon_c = \tfrac{1}{2}\sigma_{mu} \cdot \frac{V_m}{V_f} \cdot \frac{1}{E_f}$$

$$\text{i.e. } \Delta\varepsilon_c = \alpha \frac{\varepsilon_{mu}}{2}$$

$$\text{where } \alpha = E_m V_m / E_f V_f. \tag{2.5}$$

At the completion of the formation of multiple cracks at stress $E_c \cdot \varepsilon_{mu}$, further increase in stress on the composite results in fibres sliding relative to the matrix and the tangent modulus becomes $E_f V_f$. The composite fails when the stress in the fibres at a crack reaches the ultimate fibre stress. Hence the ultimate strength of the composite, σ_{cu}, is given by:

$$\sigma_{cu} = \sigma_{fu} \cdot V_f \tag{2.6}$$

and the ultimate strain, ε_{cu}, by:

$$\varepsilon_{cu} = \varepsilon_{fu} - \alpha\varepsilon_{mu}/2 \tag{2.7}$$

when the crack spacing is $2x' \cdot \varepsilon_{fu}$ is the fibre failure strain $(= \sigma_{fu}/E_f)$

The ultimate stength will be reduced if fibres are randomly aligned and/or short such that they pull out before they break. Fibre pull-out is the dominant failure mode for concrete with short, random fibres such as steel or chopped, fibrillated polypropylene.

A fibre will pull out before fibre fracture when the fibre length, l, is less than the critical fibre length l_c, defined as twice the length of fibre embedment which would cause fibre failure in a pull-out test, i.e.

$$l_c = \frac{\sigma_{fu} \cdot r}{\tau} \tag{2.8}$$

The strength of a short randomly-orientated fibre composite may be estimated from the product of the number of fibres crossing unit area of crack and the average pull-out force per fibre. If the mean pull-out length is $l/4$, then:

$$\sigma_{cu} = 2\pi r \tau N(l/4) \tag{2.9}$$

N can be estimated as:

$$N = \frac{\eta_1 V_f}{\pi r^2} \tag{2.10}$$

where η_1 is the fibre orientation efficiency factor. Values for η_1 proposed[14] are 1.0 for aligned fibres, $2/\pi$ for a random 2-D array and $\frac{1}{2}$ for a random 3-D array. Hence,

for aligned fibres, $$\sigma_{cu} = V_f \cdot \tau \frac{l}{d} \tag{2.11}$$

for random 2-D, $$\sigma_{cu} = \frac{2}{\pi} V_f \tau \frac{l}{d} \tag{2.12}$$

for random 3-D, $$\sigma_{cu} = \tfrac{1}{2} V_f \tau \frac{l}{d} \tag{2.13}$$

where d is the fibre diameter.

Thus for a 3-D short fibre concrete, if the ultimate strength σ_{cu} is to exceed the strength at which the composite cracks, then from equation (2.13):

$$V_{f\,\mathrm{crit}} > \frac{2\sigma_{mc}}{\tau(l/d)} \tag{2.14}$$

where σ_{mc} is the composite cracking stress.

For a steel-fibre concrete, for example, values of τ measured have varied, but a reasonable value of σ_{mc}/τ might be unity. The relationship between $V_{f\,\mathrm{crit}}$ and l/d for equation (2.13) is illustrated in Fig. 2.9. $V_{f\mathrm{crit}}$ is difficult to achieve in steel-fibre concrete because mixing and compaction problems increase with increasing V_f and with increasing l/d ratio.

It should be noted that the ultimate strength, σ_{cu}, will be less than the value $\sigma_{fu} \cdot V_f$ even when the fibre length is greater than the critical length since a proportion, l_c/l, of fibres will have one end less than $l_c/2$ from a crack, and will therefore pull out instead of breaking. The average stress in the fibres that pull out is $\sigma_{fu}/2$ and hence the ultimate tensile strength will be reduced to, for the aligned fibre case,

$$\sigma_{cu} = \left(1 - \frac{l_c}{2l}\right) \sigma_{fu} V_f \tag{2.15}$$

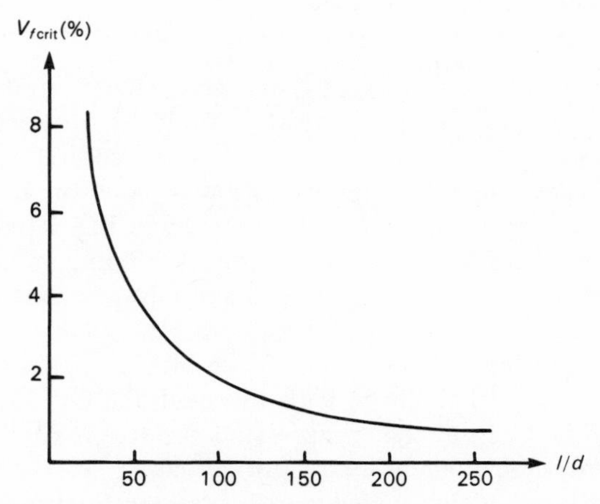

Figure 2.9 Critical fibre volume fraction, $V_{f\mathrm{crit}}$ against fibre aspect ratio, l/d, from equation (2.14) for $(\sigma_{mc}/\tau) = 1$

When $l = l_c$, equations (2.15) and (2.11) reduce to

$$\sigma_{cu} = \sigma_{fu} V_f / 2 \tag{2.16}$$

It should be emphasized that equations (2.2) to (2.16) are based on several simplifying assumptions, and the realities of composite behaviour and the practicalities of fabrication are likely to produce wide scatter in test results compared to the predicted values. These factors should be borne in mind when composite materials are designed to carry stresses.

2.2.4 *Flexural behaviour*

The presence of fibre reinforcement in flexural members may enhance properties of the unreinforced matrix in a similar fashion to the tensile case. Thus, the load at which matrix cracking is observed, the load and deflection and failure and the toughness (measured by the area under the load-deflection curve) may all be increased.

There are two further points which are important to note in considering flexural behaviour.

(i) As for plain concrete, flexural strength is generally expressed as a surface stress calculated assuming elastic behaviour with the neutral axis at mid-depth. The stress so calculated from the ultimate load is commonly termed the modulus of rupture (MOR). Since prior to failure, the neutral axis moves towards the compression surface as cracking propagates in the tensile zone, the modulus of rupture is a nominal stress which, as calculated, may be up to three times the direct tensile strength of the material.

(ii) Because an increase in moment can be accommodated by a movement of the neutral axis, flexural strengthening may occur even if, in direct tension, there is no increase in strength after the matrix cracks.

Several theories have been proposed to evaluate the flexural tensile strength of fibre-cement composites, and the reader should refer to specialist literature for the wide-ranging theoretical predictions on this subject (see, for example, references 8 and 9). The following discussion is based on a simplified stress distribution proposed by Hannant[17] and shown in Fig. 2.10(*a*). It illustrates the method to be adopted for the calculations depending on the type of stress distribution assumed and other simplified assumptions made in formulating the equations.

For the ultimate flexural behaviour shown in Fig. 2.10(*a*), the stress in the tensile zone is assumed uniform, with the neutral axis positioned at one-quarter of the depth from the compression surface. It can be shown that, provided the post-cracking tensile strength exceeds 41% of the matrix cracking strength (Fig. 2.10(*b*)), and provided there is adequate ductility in tension, then flexural strengthening can occur. By an extension of this argument, the critical

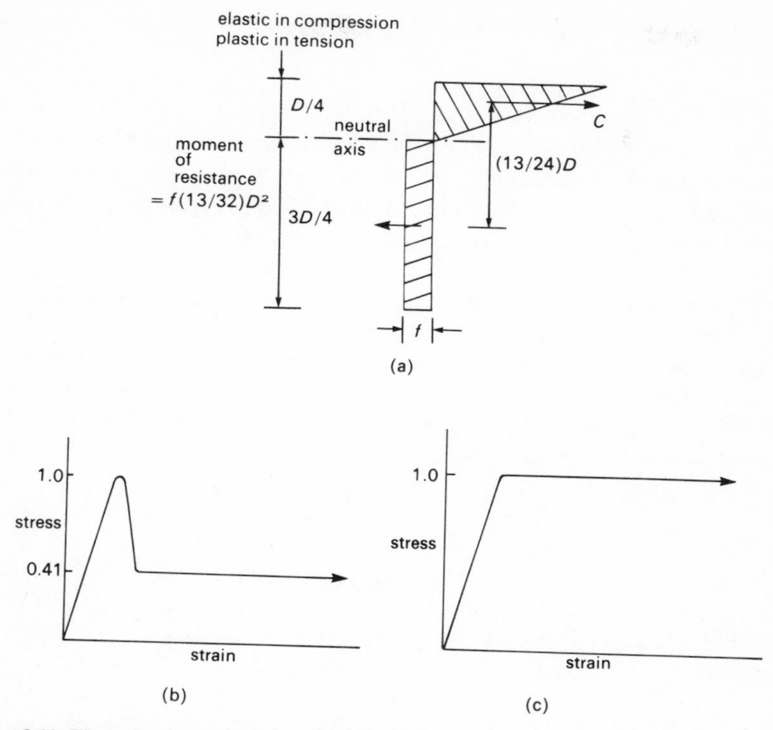

Figure 2.10 Flexural strengthening: (*a*) assumed stress distribution at ultimate load; (*b*) stress–strain curve in uniaxial tension for no decrease in flexural load capacity after cracking; (*c*) stress–strain curve in uniaxial tension yielding modulus of rupture relationships in equations (2.17)–(2.19)[17]

fibre volume for flexural strengthening is 41% of the critical fibre volume for direct tension.

As for tensile strength, the apparent modulus of rupture, σ_{MR}, for a fibre composite failing by fibre pull-out rather than fracture, is a function of $(V_f \tau l/d)$ and the orientation of the fibres. Hannant[17] has proposed the following relationships for a composite with a tensile stress–strain curve similar to that shown in Fig. 2.10(*c*):

$$\text{1-D:} \quad \sigma_{MR} \simeq 2.44 V_f \tau (l/d) \tag{2.17}$$

$$\text{2-D:} \quad \sigma_{MR} \simeq 1.55 V_f \tau (l/d) \tag{2.18}$$

$$\text{3-D:} \quad \sigma_{MR} \simeq 1.22 V_f \tau (l/d) \tag{2.19}$$

Thus, the critical factors affecting the modulus of rupture for composites in which the fibres pull out, rather than break, are the volume, shape and orientation of the fibres and the bond strength between fibre and matrix.

2.3 Steel-fibre reinforced concrete

2.3.1 *Steel fibres*

A number of fibre types are available as reinforcement. Round steel fibres, the earliest examples, are produced by cutting round wire into short lengths. Typical diameters lie in the range 0.25–0.75 mm. Steel fibres having a rectangular cross-section are produced from slitting sheets about 0.25 mm thick, and may be produced cheaply if suitable scrap steel is readily available. The melt extract process, illustrated in Fig. 2.11, is capable of producing low cost stainless and carbon steel fibres from scrap steel in many forms[18].

Attempts by manufacturers to improve the mechanical bond between fibre and matrix have resulted in the production of indented, crimped, machined and hook-ended fibres. Length/diameter or aspect ratios of fibres which have been employed vary about 30 to 250.

2.3.2 *Properties of steel-fibre concrete in the fresh state*

A statisfactory fibre concrete in the hardened state requires that the fibre reinforcement is uniformly distributed and that the concrete is well compacted. In adding fibres during the mixing, it is essential that clumps of tightly-bound fibres are broken up before entering the mix[9,19,20]. For bulk steel-fibre mixes a recommended mixing sequence is to blend fibre and aggregate before charging the mixer, by, for example, combining fibre and aggregate on a conveyor belt or chute.

The ease with which fibre concrete can be compacted during manufacture depends on the nature and amount of the fibre used and, most importantly for short fibres, on their aspect ratio[9,19,20]. The slump test has been judged to be a poor indicator of relative workability of steel-fibre concretes, since

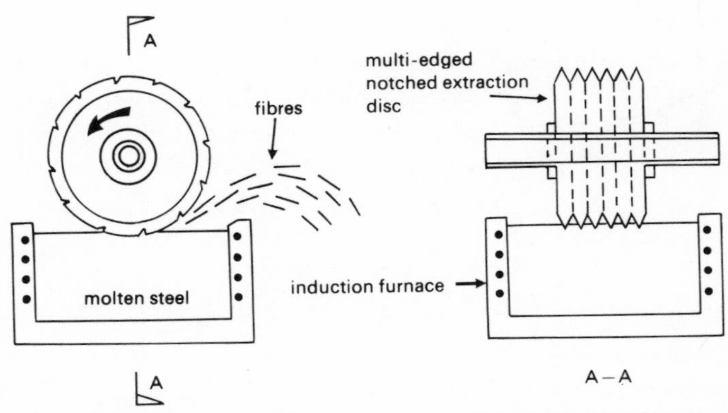

Figure 2.11 The melt extract process for the production of steel fibres

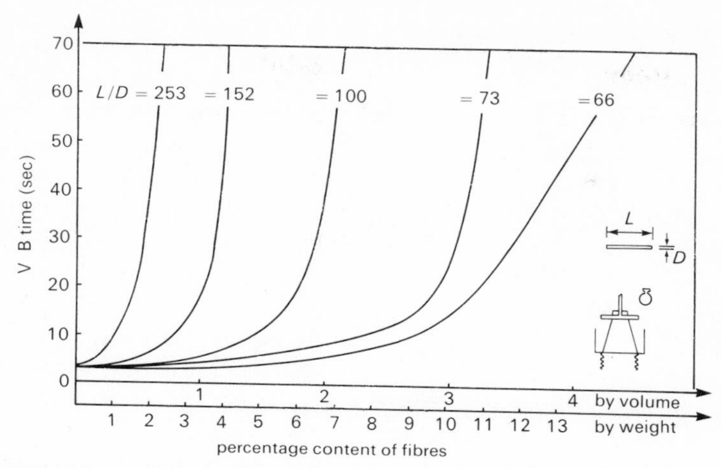

Figure 2.12 Effect of fibre aspect ratio on V-B time of steel-fibre reinforced mortar[11]

the addition of fibres to the mix changes the slump out of proportion to the workability change. The V-B test incorporates the effects of vibration and has been found to give a realistic assessment of the workability of fibre concretes[11,19].

Typical relationships between V-B time and fibre content and aspect ratio for fibre-reinforced *mortars* are shown in Fig. 2.12, which indicates that the workability of a mix decreases with increase in fibre concentration and aspect ratio. There is a critical fibre content for each aspect ratio beyond which the response to vibration rapidly decreases. Figure 2.13 shows that a reduction of maximum aggregate size facilitates the introduction of fibres, although little is gained by reducing below a 5-mm aggregate size. The use of pulverized fuel ash

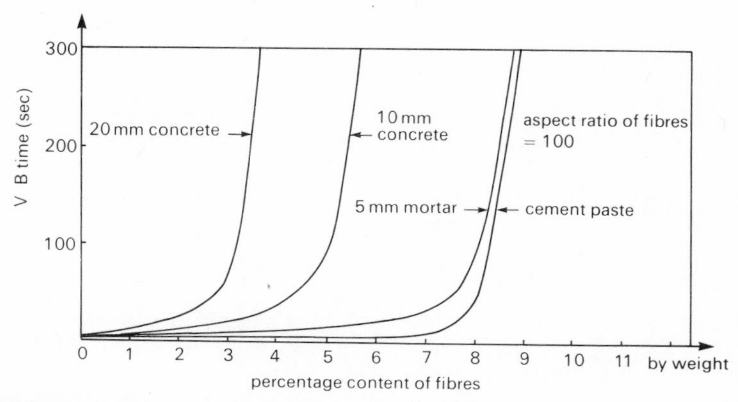

Figure 2.13 Workability against steel fibre content for matrices with different maximum aggregate size[11]

as a partial replacement of cement (30% by weight of cement) and a water-reducing admixture has been proposed[21] and adopted[22] to ease compaction problems.

The major limitation of the V-B test is that the apparatus is not thought to be convenient for field use and so it is U.S. practice[23] to use the inverted slump cone test. This measures the time to empty the steel fibre concrete mix from an inverted slump cone resting 75 mm above the bottom of a one-cubic-foot yield bucket, after a 25–30 mm diameter vibrator probe has been inserted. The probe is allowed to fall to and touch the bottom of the bucket. Test results in the range of about 11 to 28 seconds indicate a steel-fibre concrete of good workability[24].

2.3.3 *Basic properties of steel fibre concrete in the hardened state*

Direct tensile behaviour. The uses of steel fibres have mainly been in conjunction with concrete or mortar and there has been little emphasis placed upon the production of thin sheet products. The practical limitations imposed on the volume and aspect ratio (l/d) of fibres for ease of mixing mean that the direct tensile behaviour is typified by Fig. 2.6(b). Tensile stress–strain curves of this type for 1.73% by volume of steel fibres in mortar are shown in Fig. 2.14[25]. There is some modest increase in tensile strength due to fibre reinforcement, but more substantial is the increase in toughness as measured by the area under the stress–strain curve.

Compressive behaviour. It has generally been found that the presence of fibres in concrete produces no or only modest increase in compressive strength, although the increased ductility resulting from fibre addition may be advantageous, particularly in over-reinforced concrete beams where a brittle failure can be changed into a ductile one[26].

Flexural behaviour. The major factor influencing the flexural strength (charac-

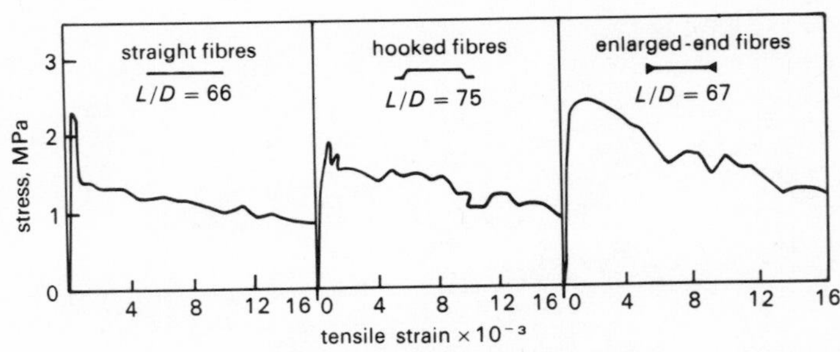

Figure 2.14 Stress–strain curves for steel-fibre reinforced mortars in tension (1.73% fibre volume)[25]

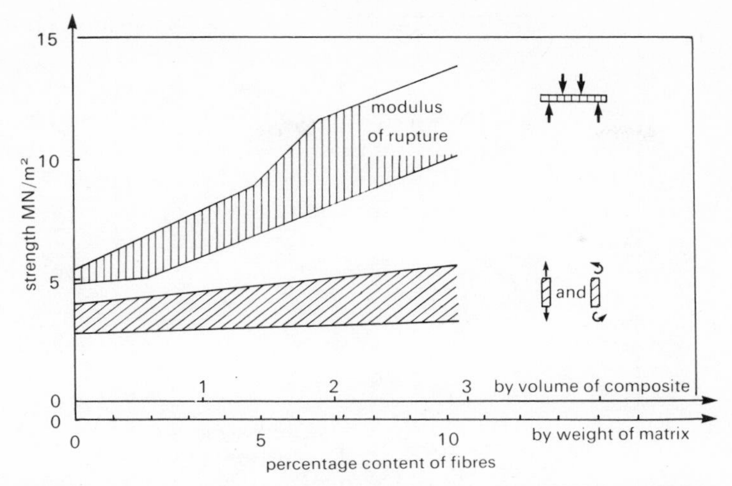

Figure 2.15 Flexural, direct tensile and torsional strengths of steel-fibre reinforced mortar and concrete[11]

terized by the modulus of rupture) has been found to be the term $V_f l/d$, a conclusion which supports the theoretical relationships in equations (2.17) to (2.19). In Fig. 2.15, the improvements in flexural strength for mortar and concrete with a range of fibre l/d ratios are shown (together with more modest increases in tensile and torsional strength)[11]. A survey of data by Johnston for flexural strength[27] is shown in Fig. 2.16 in which the increase in $V_f l/d$ from 40 to 120 (a practical limit for workability considerations) increases the flexural strength by about 25%. The fatigue flexural strength after 10^5 cycles is increased by a similar proportion.

Toughness and impact strength. The area under the complete load–deflection curve (or under a prescribed part of the curve) can be described as a measure of the toughness or energy absorption capability of the material. The results of data[28] for the area up to the maximum stress on the load–deflection curve are shown in Fig. 2.16, indicating the substantial energy-absorbing capability (compared to plain concrete) of the fibre-reinforced material.

Improvements in impact strength for fibre-reinforced concretes are highly dependent on the type of fibre and the method of test[23,29–31]. This is illustrated in Fig. 2.16 which contains results from a falling weight method and a pendulum-type impact machine. The results are also dependent upon the type of fibre, with crimped fibres yielding greater improvement in impact strength than straight fibres[31].

Durability of steel-fibre concrete. As for conventional reinforced concrete, steel fibres will be protected from corrosion provided the alkalinity of the matrix is maintained in the vicinity of the fibres. Carbonation of the concrete

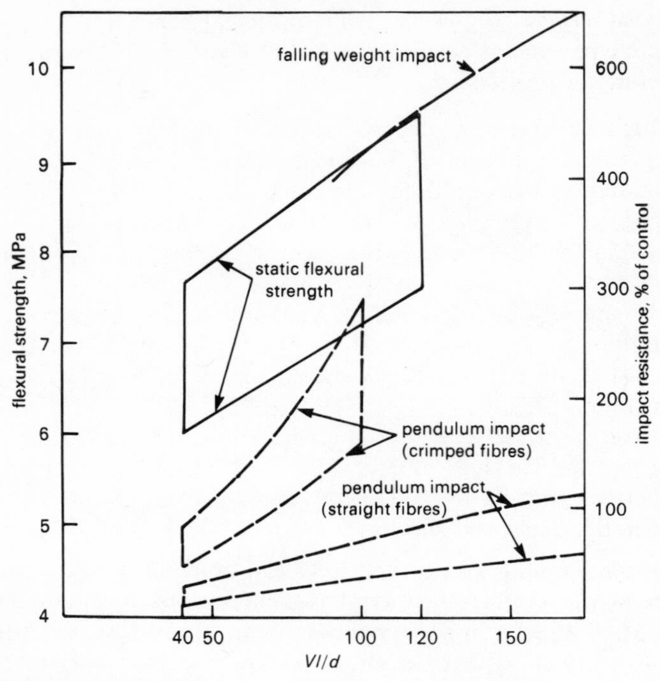

Figure 2.16 Influence of fibre volume fraction, V_f, (%) and aspect ratio, l/d, on flexural properties of steel-fibre reinforced concrete[27,28]

matrix can lead to corrosion of the fibres and any deterioration may be accelerated if the concrete is cracked. Since the failure mechanism of the steel fibre concrete is by fibre pull-out rather than fibre fracture, a considerable reduction in fibre diameter due to corrosion could conceivably be tolerated since the uncorroded fibre strength is not fully utilized (provided, of course, that corrosion does not reduce the interfacial bond strength).

For freeze-thaw durability, proper air-entrainment is just as necessary for steel-fibre concrete as for plain concrete.

2.3.4 *Applications of steel-fibre concrete*

Applications of steel-fibre concrete may be classified into six categories[22,30,32–34]:

(a) highway and airfield pavements
(b) hydraulic structures
(c) fibrous shotcrete
(d) refractory concrete
(e) miscellaneous precast applications
(f) structural applications.

Highway and airfield pavements. Three principal benefits have led to steel-fibre concrete pavements being selected as the best engineering solution on a number of major projects.

(i) A high flexural strength (compared to plain concrete) results in a reduction in the required pavement thickness for a given design life, or an extended design life for the same thickness. By a rule of thumb approach[35], if the flexural strength of fibre concrete is twice that of plain concrete, then the required thickness of the former is about 70% of the latter.

(ii) The transverse and longitudinal joint spacing may be increased. Under conditions of restrained shrinkage, the greater tensile strain capacity of steel fibre concrete results in lower maximum crack widths than in plain concrete[32–34].

(iii) The resistance to impact and repeated loading is increased. Mixing, placing and finishing of the fibre concrete is very similar to plain concrete, and slabs have been laid using semi-manual methods, concreting train and slip-form paver[32–34] (Fig. 2.17).

Projects have involved new pavement construction or repair to existing pavements by the use of overlays which may be bonded or unbonded to the slab beneath[35]. At McCarran International Airport, Nevada, USA[36], both types of construction have been employed, the details of which are given in Table 2.2.

Figure 2.17 CPP 60 slip form paver laying steel-fibre reinforced concrete as an overlay trial on the M10 motorway, England[35]

Table 2.2 Steel fibre concrete in paving at McCarran International Airport, Nevada.[22,36]

				Fibres		*Binder (kg/m²)*		
Date	*Type of application*	*Volume (m^3)*	*Depth (mm)*	*Dimensions (mm)*	*(kg/m^3)*	*cement*	*flyash*	*Flexural strength (MN/m^2 days)*
March 1976	Unbonded overlay	8410	152	25 × 0.25 × 0.55 (slit sheet)	95	356	148	6.8 (28) 7.6 (90)
1979	slab on grade	12950	178	50 × 0.5 dia. (hooked ends)	50	385*	149*	7.2 (28)

*Other constituents kg/m³: coarse aggregate 780
sand 810
water 203

Hydraulic structures. The principal reason for using steel-fibre concrete in hydraulic projects is its resistance to cavitation/erosion damage by high-velocity water flow. The success of steel-fibre concrete exposed to erosion/cavitation forces has been demonstrated in laboratory tests, where the fibre concrete has lasted three times as long as plain concrete when exposed to a water velocity of 37 m/s[30].

A recent, notable example of the use of steel-fibre concrete to combat a severe cavitation/erosion problem was the repair of a spilling basin undertaken at Tarbela Dam, Pakistan, involving the placement of over 3000 m³ (in a 500 mm layer) of concrete containing about 1% by volume of 25 × 0.25 × 0.55 mm slit-steel fibres. Further work at Tarbela has involved over 3000 m³ of steel fibre concrete in a repair to a plunge pool.

Fibre shotcrete. Fibre shotcrete has been used in rock slope stabilization, tunnel lining (Fig. 2.18) and bridge repair. A thin coating of plain shotcrete, applied monolithically on top of the fibre shotcrete, may be used to prevent surface staining due to rusting. Traditional sprayed concrete techniques may be modified to include fibre mixing with the wet or dry constituents. A technique developed by the Besab Company in Sweden, shown in Fig. 2.19, involves the pneumatic conveying of fibres from a rotary fibre feeder to a nozzle via a 75 mm diameter flexural hose[18]. The equipment can be used with almost any standard dry shotcrete machine and is capable of handling fibres with aspect ratios up to 125.

In certain applications, fibre shotcrete may warrant consideration as an economic alternative to traditional construction for the following reasons.

(i) The cutting, bending and fixing of conventional reinforcement to curved or rough surfaces is eliminated.

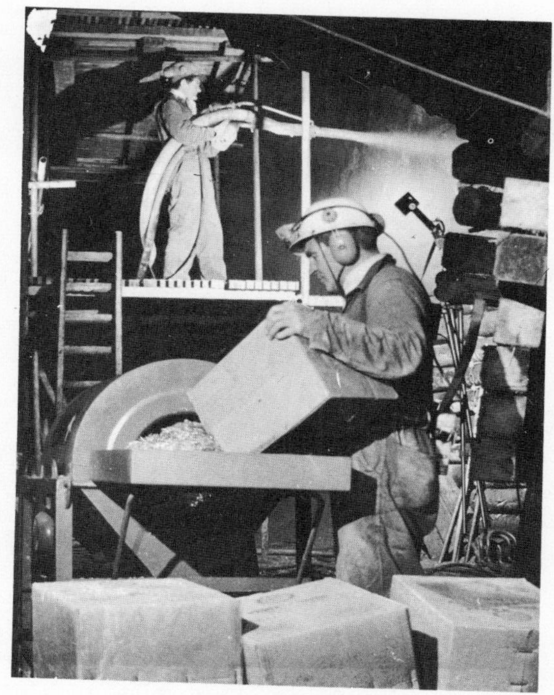

Figure 2.18 Shotcreting steel fibre-reinforced concrete using BESAB technique

(ii) The time required to apply a lining is significantly reduced, which may suit applications where freshly-exposed rock surfaces need rapid sealing.
(iii) The material can be applied in thin sections to follow closely the surface contours, with significant material savings.

The reinforcing effect of fibres in shotcrete applications can be at least as good as that obtained with more conventional placing techniques, mainly because the fibres are preferentially aligned in two dimensions by the mode of application of relatively thin coating. The strength of the concrete matrix can benefit from the low water: cement ratio which may be used for shotcrete applications[33,34].

A novel example of the use of fibre shotcrete has been in the protection of the structural steelwork of the Hong Kong and Shanghai Bank building in Hong Kong. A 14 mm layer of polymer-modified mortar, reinforced with melt-extract steel fibres and applied by spraying, has formed part of the protective treatment of steel elements in the huge support structure.

Refractory concrete. Steel-fibre reinforced refractory concretes have shown themselves to be more durable than their unreinforced counterparts when

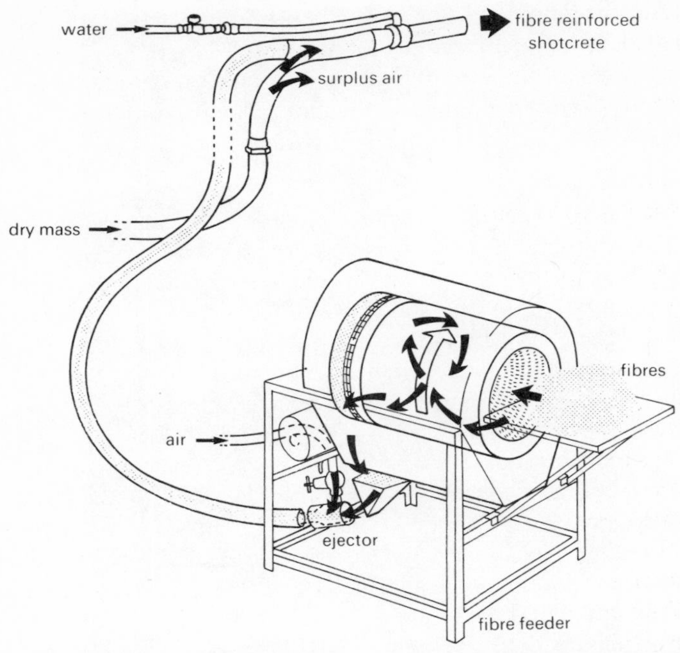

Figure 2.19 Fibre shotcreting equipment developed by the BESAB company

exposed to high thermal stress, thermal cycling, thermal shock or mechanical abuse[18,32]. The increased service life is probably due to a combination of crack control, enhanced toughness and the spall and abrasion resistance imparted by the steel fibres. The availability of lower-cost, more durable, stainless steel fibres from the melt extract process has extended the use of fibre-reinforced refractory concrete.

Specific applications have included stirring devices, lances and plunging bells for the desulphurization of iron, doors and linings for furnaces and coke ovens and roofs for electric arc furnaces. Through use of the shotcrete technique, the material has also been used for lining ash hoppers and lining a flame exhaust duct at a missile launch complex at Cape Canaveral.

Miscellaneous precast applications. A variety of applications have made use of the improved flexural and impact strengths and resistance to handling misuse of steel fibre concrete. Examples include manhole covers, concrete pipe, burial vaults and machine bases and frames.

Structural applications of steel-fibre reinforcement. Structural applications where fibres alone provide the reinforcement are rare, although one notable example consisted of precast slabs about 1.1 m square and 65 mm thick, supported by a tubular steel space frame for a demountable car park at

London Airport. Three per cent by weight of 0.25 mm diameter by 25 mm long fibres provided the reinforcement for the slabs[37].

The prospects for the uses of fibre reinforcement in structural applications have recently been reviewed by Swamy[38], who outlines the following possibilities.

(i) Fibre reinforcement can inhibit crack growth and crack widening, which may permit the use of high-strength steels without excessive crack widths or deformation at service loads.
(ii) Fibre reinforcement can act effectively as shear reinforcement. Punching shear strength of slabs may be increased and a sudden punching shear failure transformed into a slow ductile one.
(iii) In prestressed concrete members, the addition of fibre reinforcement has been found to reduce transmission lengths and prestress losses due to elastic shortening, shrinkage and creep.
(iv) Fibre reinforcement can provide increased impact strength for conventionally reinforced beams. Resistance to local damage and spalling is also increased.
(v) Fibre reinforcement can provide enhanced stability and integrity to preserve conventionally-reinforced concrete structures subject to earthquake and explosive loading.

Although extensive data are now available to design reinforced and prestressed concrete structural elements incorporating steel fibres, there has been as yet only limited use of such fibres in actual structures[33,34], but there is considerable scope for future applications in this respect.

2.4 Glass-fibre reinforced cement (grc)

The major efforts in utilizing glass fibres has been concentrated upon the reinforcement of cement or mortar matrices in the development of thin-sheet products[39]. Relatively little attention has been paid to glass-fibre reinforced concrete, although applications, principally as pipes, have arisen.

2.4.1 *Glass fibres*

Many varieties of glass fibres available for different applications, such as 'E' glass used in the reinforcement of plastics, have inadequate resistance to the alkalis present in portland cements. As discussed previously, Dr A.J. Majumdar at the Building Research Establishment showed that it was possible to melt and, more importantly, to fiberize a glass composition which, by virtue of containing about 16% of zirconium oxide, had improved alkali-resistant characteristics. The composition of 'E' and Alkali-Resistant (AR) glass is given in Table 2.3.

In the production of glass fibres, molten glass passes from a furnace into a series of tanks, called bushings, each bushing having many hundreds of small

Table 2.3 Compositions of glass fibres[39]

Component	*E-glass (%)*	*Alkali-resistant glass (%)*
SiO_2	55	71
Al_2O_3	15	1
B_2O_3	7	—
CaO	21	—
Alkalis (Na_2O, K_2O etc)	2	11
ZrO_2	—	16
LiO_2	—	1

nozzles in its base from which the molten glass exudes and is drawn into filaments 10–20 μm in diameter. A protecting lubricating size is transferred on to the filaments before they are gathered into bundles called strands, each containing about 200 filaments, and wound on to cardboard tubes.

Strands may be combined to form a roving or woven in various types of mat or cloth.

2.4.2 *Production techniques of glass-fibre reinforced cement*

The constituents of a typical grc mix are given in Table 2.4. A variety of admixtures has proved useful in grc manufacture. Polymers are not always used in mixes but, when added, are intended to improve some physical property such as moisture movement.

The processes of manufacture of grc products may be classified under three headings:

(i) spraying
(ii) premix
(iii) incorporation of continuous rovings.

In the *spray* process, the glass-fibre strand is chopped into lengths between 10

Table 2.4 The constituents of a typical grc mix[39]

	Weight %
Cement	38.9
Sand	38.9
Water	12.8
Fibre	5.0
Admixture	0.4
Polymer	4.0
	100.0

and 50 mm and blown in a spray simultaneously with the mortar slurry on to a mould or flat bed. In the widely used spray-suction process, excess water is removed by vacuum.

With *pre-mixing*, short strands (usually 25 mm in length) are mixed into the mortar paste or slurry before further processing by casting into open moulds, pumping into closed moulds, extruding or pressing. Care must be taken in the mixing process to avoid fibres tangling and matting together and to minimize fibre damage.

Continuous rovings are impregnated with cement slurry by passing them through a cement bath before they are wound on to an appropriate mandrel. Additional slurry and chopped fibres can be sprayed on to the mandrel and compaction can be achieved by the application of roller pressure combined with suction. Fibre volumes in excess of 15% have been achieved with this process, which has application in pipe manufacture.

Research into the replacement of asbestos fibres with glass fibres has led to production of flat and corrugated grc sheet based on the Hatschek process.

2.4.3 *Properties of glass-fibre reinforced cement*

The main emphasis in research and practical applications has been placed upon the behaviour of AR fibres in an OPC matrix and the discussion in this section is limited to this matrix. Other matrix formulations are discussed subsequently. Most of the property data published has been from tests on sheets manufactured by the spray-suction process.

Direct tensile behaviour. The 28-day tensile properties of grc generally conform to the behaviour typified by the stress–strain relationship in Fig. 2.6(*c*), with the occurrence of multiple cracking of the matrix. The effects of varying the fibre length and fibre content on the 28-day properties are shown in Fig. 2.20(*a*) and (*b*) for two storage conditions[40]. Increase in fibre volume percentage results in an increase in the apparent matrix cracking stress in line with the theoretical predictions (see Fig. 2.5). The greatest benefit derived from longer fibres is the improvement in the ultimate failure stress and strain.

The long-term properties of grc have been studied up to 10 years and a summary of the results is given in Table 2.5 (which also includes flexural, modulus and impact test results)[41]. In relatively dry air storage conditions the tensile strength shows little change, but in wet environments, which include natural weathering, significant reductions in strength take place. The strength reduction is accompanied, more significantly, by a reduction in the strain to failure and the material may become brittle on ageing. The tensile stress–strain curves of grc composites containing 4.4 and 8.2 volume percent of 40 mm long fibres and aged for 5 years in three different environments are shown in Fig. 2.21. The matrix has gained in strength with age in the moist environments

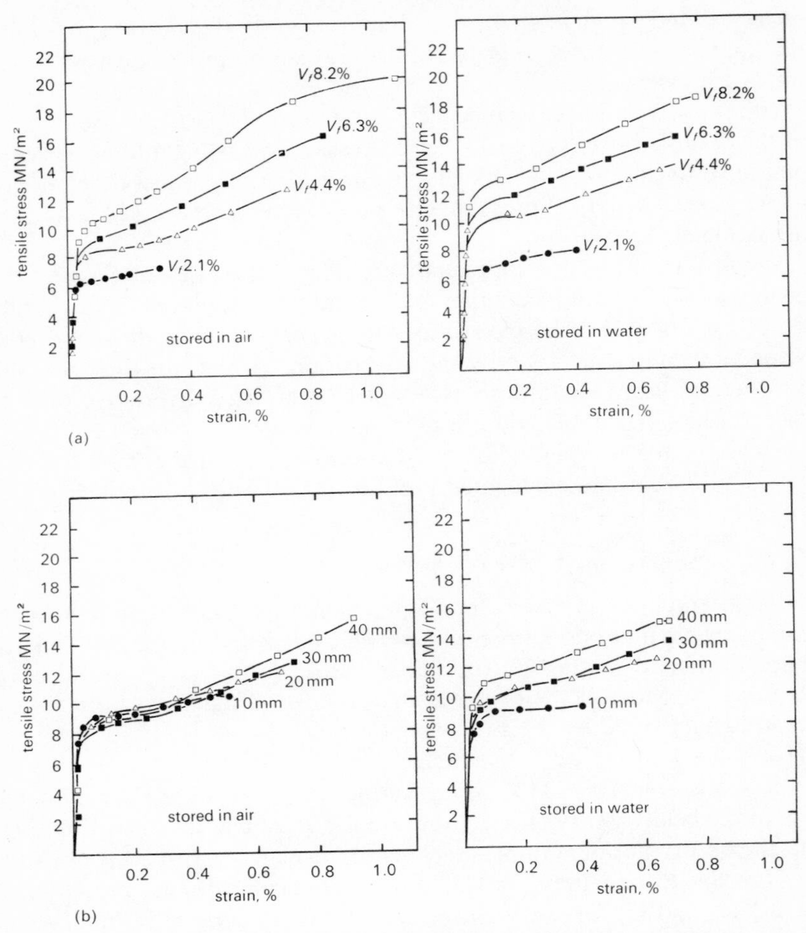

Figure 2.20 Tensile stress–strain curve of grc composites at 28 days for storage in air and in water: (*a*) 30 mm long fibres with different fibre volume fractions (*b*) 4 volume % fibres with different fibre lengths[40]

but the strain to failure of the composite is close to that of the matrix alone, except for the higher glass-content material which has retained a significant proportion of its pseudo-ductility after prolonged weathering.

Ageing of grc. The mechanisms governing the changes with age and conditions of use are not yet fully understood, but there appear to be contributions from three areas[42].

(i) *Hydroxyl attack on glass fibres*

The reduction in strength of grc with time under wet conditions is considered to be primarily due to the decrease in the fibre-strand

Table 2.5 Measured mean strength properties of spray dewatered OPC/grc at various ages (5 wt per cent glass fibre)[41]

	Total range for air and water storage conditions at 28 days	1 year			5 years			10 years		
		*Air**	*Water⁺*	*Weathering*	*Air**	*Water⁺*	*Weathering*	*Air**	*Water⁺*	*Weathering*
(a) Bending										
MOR (MN/m^2)	35–50	35–40	22–25	30–36	30–35	21–25	21–23	31–39	17–18	15–19
LOP (MN/m^2)	14–17	9–13	16–19	14–17	10–12	16–19	15–18	14–16	16–17	13–16
(b) Tensile										
UTS (MN/m^2)	14–17	14–16	9–12	11–14	13–15	9–12	7–8	11–15	6–8	7–8
BOP (MN/m^2)	9–10	7–8	9–11	9–10	7–8	7–9	7–8	9–10	6–8	6–8
Young's modulus (GN/m^2)	20–25	20–25	28–34	20–25	20–25	28–34	25–32	25–33	25–31	27–30
(c) Impact strength										
Izod Nmm/mm^2	17–31	18–25	8–10	13–16	18–21	4–6	4–7	15–22	2–3	2–6

*At 40 per cent relative humidity and 20°C
⁺At 18–20°C

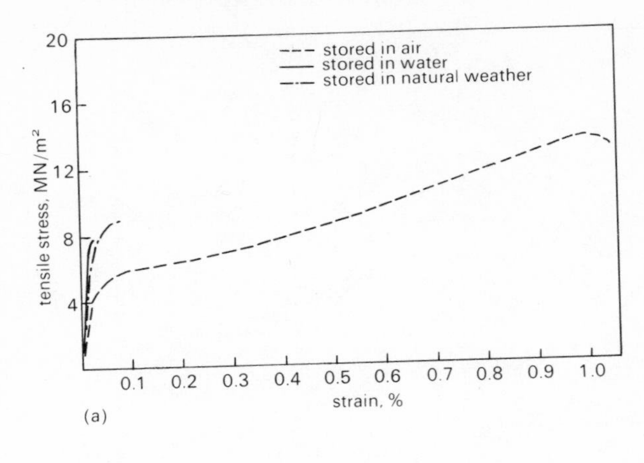

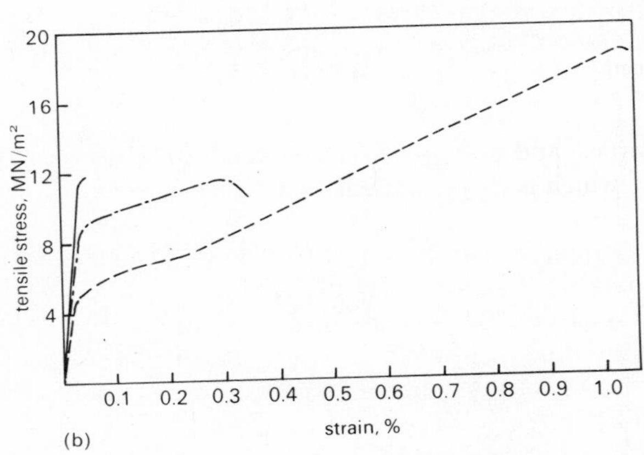

Figure 2.21 Tensile stress–strain curves of grc composites at age 5 years (*a*) 4.4 volume % (*b*) 8.2 volume % of 40 mm long fibres[41]

strength caused by the attack of the hydroxyl ions from the highly alkaline cement pastes.

(ii) *Densification of the fibre/matrix interface due to hydration products*
The fibre/matrix interface of specimens stored for 5 years in water has been found to be very dense and to have a higher contact area than unaged or oven dry samples. This will contribute to a greatly increased frictional bond and a hard, stiff material will result at points where fibres bend as they align themselves with the direction of stress across a crack. Thus fibres may fracture at cracks with little fibre pull-out.

(iii) *The formation of calcium hydroxide*
Evidence has been found of the formation of calcium hydroxide in and around glass-fibre bundles. This will contribute to the densification of

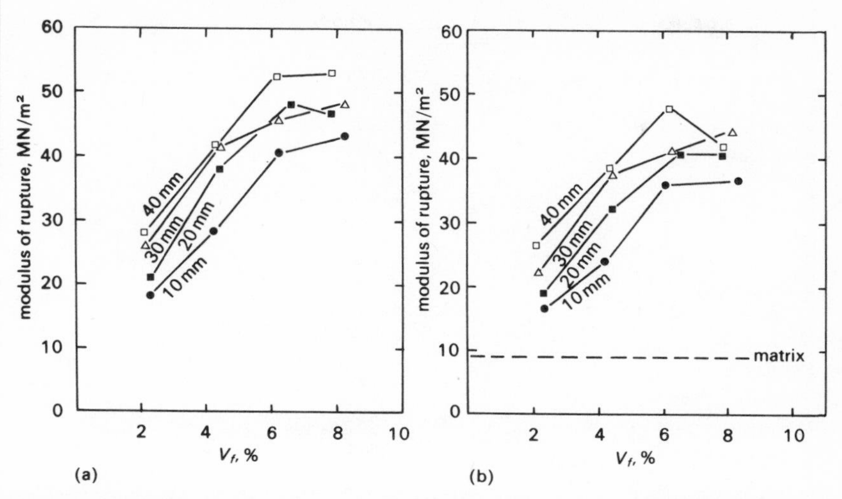

Figure 2.22 Relation between fibre volume fraction and modulus of rupture of grc at 28 days for different fibre lengths: (*a*) stored in air; (*b*) stored in water[40]

the interface and may be of further significance to grc durability in a manner which is as yet unclear.

Properties in flexure. Flexural strengths of Cem-FIL glass-reinforced cement show similar trends to tensile strength for variations in fibre length or volume and age and storage conditions. Relationships at 28 days between modulus of rupture and fibre volume for different fibre lengths are shown in Fig. 2.22. Table 2.5 indicates that the flexural strengths of waterstored and weathered specimens reduce in time and approach the flexural strength of the matrix alone.

Toughness and impact strength. As Fig. 2.20 and 2.21 illustrate, the area under the tensile stress–strain curve (or the energy absorbed to failure) at 28 days or after 5 years' air storage is substantial, i.e. over 100 kJ/m^3 compared to less than 5 kJ/m^3 for the matrix alone. Thus new sheets may readily withstand impacts arising during handling and fixing. With time, however, in natural weathering, the energy absorbed reduces to about 5 kJ/m^3 and this may present problems if sheets are subject to impact in service. A similar reduction in impact strength is obtained from Izod tests on aged water storage and weathered specimens, as shown in Table 2.5.

2.4.4 *Applications of glass-fibre reinforced cement*

GRC has proved to be an attractive material for a wide range of applications. To date, primary load-bearing or fully structural applications (such as floor

Table 2.6 Applications of glass-fibre reinforced cement[39]

The building industry	Architectural cladding and sandwich panels Formwork systems Frameless housing systems based upon GRC sandwich panel construction Fencing systems including noise barriers Ducting Roofing elements
The Water industry	Storage tanks Pipes, flumes and raceways Sewer lining elements Swimming pools
The fire protection industry	Fire stop partitioning and cladding systems
The marine industry	Floating pontoons and buoys Still-water boats
The agricultural industry	Trough units Tanks and drainage elements

(*a*)

(b)

(c)

Figure 2.23 Examples of the use of grc. (*a*), Cladding panels; (*b*), large-diameter pipe; (*c*), bridge deck form work

slabs, beams, columns, etc., failure in which could result in direct loss of life or limb) have been limited. However, as experience with the material has developed certain quasi-structural applications, e.g. transient load-bearing bridge deck formwork, have been permitted. The products in Table 2.6, which is not exhaustive, with examples in Fig. 2.23, are either being currently supplied to contract or are in an advanced field-trial stage of market development[39].

An example of a development as a load-bearing application was the use of grc elements for the fabrication of a temporary shell structure for a flower pavilion in Stuttgart (Fig. 2.24), in which the 10 mm thick grc results in a considerable saving in structural weight, compared to the 40–50 mm thick reinforced concrete structure which might otherwise have been necessary.

More attention has been paid to the philosophy of design of elements in grc than to other fibre-reinforced concretes. Reference 39 provides a detailed insight into design approaches for grc elements. The following factors need to be considered in design.

(i) *Strength properties of grc.* In general, the limit of proportionality (LOP) in flexure is taken as the basis in design. The pseudo-ductile behaviour above the LOP provides a factor of safety against failure, but as this pseudo-ductility and the modulus of rupture may fall in wet and natural weathering conditions,

Figure 2.24 Thin shell construction in grc flower pavilion in Stuttgart

design stresses are normally well below the LOP. Design working stresses of $3MN/m^2$ (tension) and $6MN/m^2$ (bending) have been quoted[43].

(ii) *Shrinkage, moisture and thermal movements.* When these movements are restrained, tensile stresses will be induced in the grc which may be of the same order or higher than those resulting from applied loads.

It should be noted that a number of problems are appearing associated with grc cladding in the U.K. In particular, certain configurations of sandwich panels may eventually crack as a result of restrained shrinkage, moisture and thermal movements. Design practice for such panels may need revision in the light of current investigations.

(iii) *Manufacturing process.* It must be recognized that differences will exist between laboratory-produced material and that produced by the manufacturer. Furthermore, there will be variations in material quality within the manufacturing process. For example, there will be inherent variations in fibre orientation, dispersion and in curing effects; small, but for thin sections, significant thickness variations may occur which must be acknowledged in the design approach.

(iv) *Behaviour of full-size components.* Care must be taken in the transfer of data obtained from small-size test coupons in bending to its use in the design of a full-size component.

(v) *Interaction with associated materials.* These will include infills for sandwich construction, surface coatings, fixings and jointing materials.

2.4.5 *Recent developments in glass-fibre reinforced cement*

A second generation of AR alkali-resistant glass fibres (called Cem-FIL 2) has been developed to improve the long-term strength and strain capacity of grc. Accelerated tests on specimens containing these fibres have indicated improved composite behaviour. For example, Fig. 2.25 compares the retention of strain to failure in bending of composites containing Cem-FIL 2 and the original Cem-FIL 1 AR fibres. The improvement is achieved by a surface treatment of the fibres, the nature of which has not been revealed, which effectively reduces the rate of reaction between the glass fibre and the cement matrix[44].

An alternative approach to achieve long-term durability of grc has been the use of 15% by volume of polymer addition to the mortar matrix. The increased matrix cost is balanced by the use of cheaper E-glass fibres[45]. The material has the trade name Forton. A coherent, comparatively soft, polymer film is apparently spread uniformly over the fibre surface, isolating the glass from the hard and brittle cement matrix. Reactions due to active acid groups in the

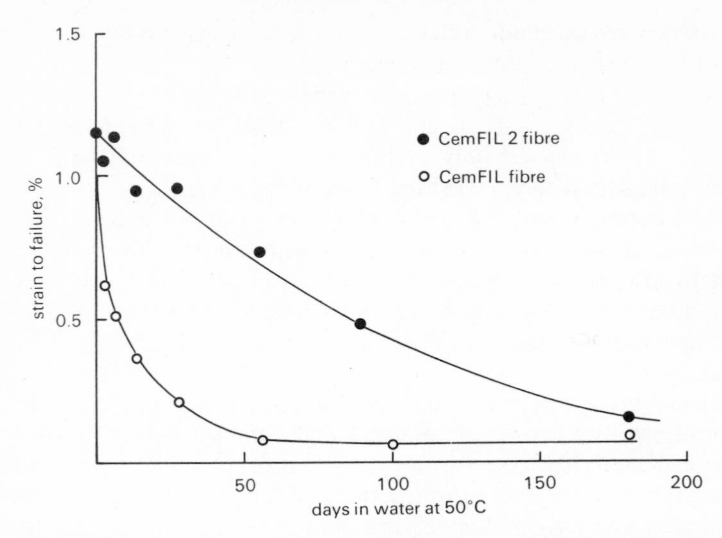

Figure 2.25 Retention of strain to failure in bending in accelerated ageing tests for grc composites containing 5% weight Cem FIL fibres[44]

polymer decrease the amount of calcium hydroxide crystals formed. Bijen[46] has concluded that after accelerated ageing for 26 weeks, the tensile strength, strain at ultimate tensile stress, bending and impact strengths of Cem-FIL 1 and Cem-FIL 2 grc have dropped below those of the polymer-modified grc, but that all the materials became comparatively brittle in the long term.

The development of Cem-FIL 2 fibres has made possible reductions in the fibre content required to achieve acceptable long-term properties for asbestos-cement replacement products, with the result that a 3% by weight fibre content sheet is apparently cost-competitive and meets the required performance specification[47]. It is interesting to note that a second fibre component (believed to be cellulose) must be added to the cement slurry if asbestos-cement process equipment is to be used. This is because the glass fibres, with much greater diameters than asbestos fibres, are unable to control the slurry drainage as required in a Hatschek type process.

2.5 Polypropylene-reinforced cement and concrete

Polypropylene fibres have been used commercially as small volume fractions of short fibres incorporated in concrete. A fibre volume of about 1% is the maximum that can be accommodated in practice, but even with a fibre content as low as 0.5%, considerable improvement in the impact strength of the hardened concrete can be achieved.

The use of higher volumes of chopped filaments or short, fibrillated film lengths of polypropylene to reinforce thin cement sections has been investigated. The greatest potential for a thin-section product, directed towards asbestos-cement replacement, may lie with the polypropylene reinforcement in the form of continuous networks of opened fibrillated film.

2.5.1 *Polypropylene fibres*

Polypropylene is one of the cheapest available polymers and this, together with a high existing production capacity likely to be adequate for a new use in cement composites, means that it is attractive as a reinforcing fibre. Polypropylene fibres are resistant to most chemicals and it would be the cementitious matrix which would deteriorate first under aggressive chemical attack. The melting point of polypropylene is high enough (165°C) that a working temperature of 100°C may be sustained for short periods without detriment to fibre properties. The fibres are combustible but this does not invalidate the potential use as a sheeting material.

Polypropylene fibres are avialable in two forms; monofilaments produced from spinnarets and, more commonly, film fibres produced by extrusion. The extruder is fitted with a die to produce a tubular or flat film, which is then slit into tapes and strectched to between 8 and 20 times its original length. This imparts a molecular orientation to the film resulting in a high tensile strength (about 400 MN/m^2) in the direction of drawing and a weakness in the lateral direction, such that it can be fibrillated or split by passing the film over a system of pins on rollers. The fibrillated film may then be:

(i) twisted into twine and chopped, usually into 25–50 mm lengths, for use in concrete (Fig. 2.26(*a*));
(ii) opened to produce continuous networks for use in thin sheet manufacture (Fig. 2.26(*b*)).

Fibrillated film may also be woven to produce flat meshes which may be used as thin cement sheet reinforcement, although the open network composite has the better properties in the hardened state.

Polypropylene fibres are hydrophobic, which can be advantageous in the mixing process since the fibres need no lengthy contact during mixing and need only be dispersed evenly through the mix, but disadvantageous to hardened properties since no physico-chemical bond between fibre and matrix is established. A mechanical bond is formed as cement paste penetrates the mesh structure between individual fibrils of the chopped length or continuous network.

2.5.2 *Polypropylene-reinforced concrete—manufacture and properties in the fresh state*

A variety of mixers have been used in practice to mix chopped polypropylene fibres into concrete. As the mixing need only disperse the fibres evenly through the mix, they are, therefore, added shortly before the end of mixing the normal

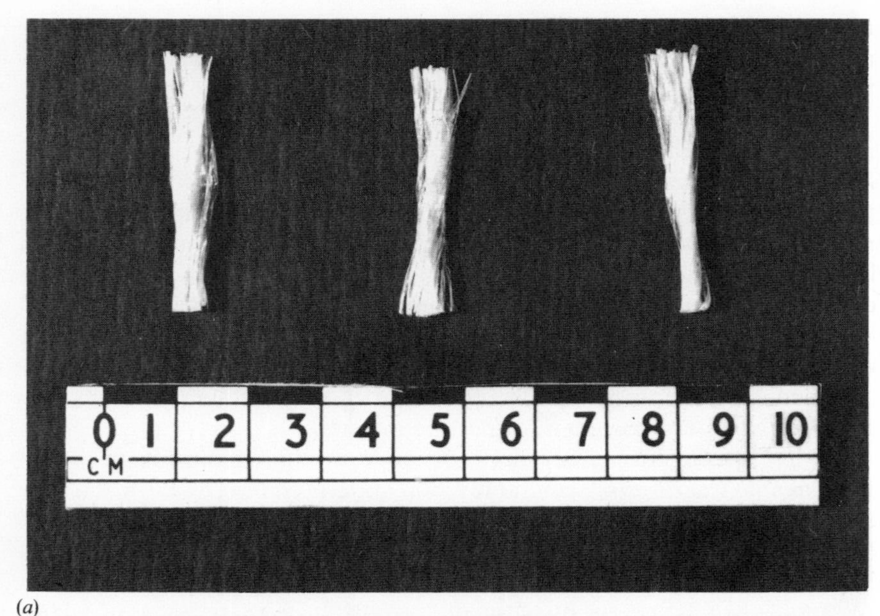

(*a*)

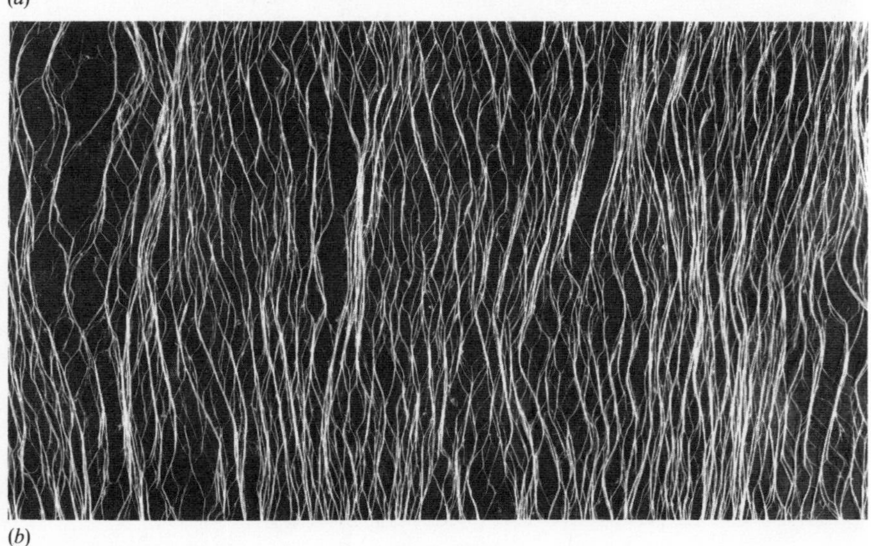

(*b*)

Figure 2.26 Polypropylene fibre reinforcement: (*a*) fibrillated film length; (*b*) opened network

constituents. Too long a mixing time may lead to undesirable shredding of the fibres.

At precast factories, continuous twine is chopped with a guillotine knife into short lengths which then drop on to a running conveyor belt taking the other mix components to the mixer. Surface coatings of polypropylene-reinforced mortar may be achieved by 'shotcreting' using normal equipment. The fibres are cut to about 20 mm lengths to enable smooth transport of the dry mix through air hoses and nozzles. Water is then added at the gun orifice.

The workability of polypropylene fibre concrete has not received the same attention in the literature as steel-fibre concrete. Work reported on normal and lightweight aggregrate concrete suggests that the compacting factor test relates most suitably to production experience[48]. A natural aggregate concrete mix of medium workability (compacting factor about 0.88) may reduce to a low workability mix (compacting factor about 0.75) following the addition of 1% of chopped 35 mm polypropylene fibres.

Interestingly, polypropylene monofilaments have been used in concrete in quantities of about 0.1–0.2% by volume to alter the rheological, rather than the hardened, properties of the material. In the early 1970s, John Laing Research and Development in the U.K. developed a fabrication technique in which highly air-entrained concrete (about 40% air by volume) is stabilized by the fibres[49]. A decorative sculptured finish may be imprinted which does not slump back as the press is removed, eliminating the need for complex moulds (Fig. 2.27).

Figure 2.27 Cladding panel production: aerated concrete mix stabilized by 0.1– 0.2% of chopped polypropylene filaments

2.5.3 *Properties of polypropylene-reinforced concrete in the hardened state*

Tensile strength. The tensile strength of concrete is essentailly unaltered by the presence of a small volume of short polypropylene fibres, although there is a residual load capacity after matrix cracking as the fibres pull out from the cracked surfaces.

Flexural strength. Similarly, changes in the flexural strength of concrete due to polypropylene fibre addition are small and generally cannot be distinguished from the normal scatter of results. Of much greater importance is the post-cracking behaviour and the ability to continue to absorb energy as fibres pull out. (Use of longer fibres may result in some fibres fracturing at the crack surfaces rather than pulling out.) Typical load–deflection curves of a polypropylene fibre concrete illustrating the post-cracking behaviour are shown in Fig. 2.28, for two types of chopped polypropylene film[50]. The test beams use were 100 × 100 × 500 mm loaded at midspan over a 400 mm span. The load–deflection curve of a plain concrete is given for comparison.

Impact strength. Using the area under the load–deflection curve in slow flexure as a measure of the ability to absorb flexural impact energy, Table 2.7 indicates the energy absorbed up to a central deflection of 10 mm. In addition to the polypropylene-reinforced specimens, results for the plain matrix and

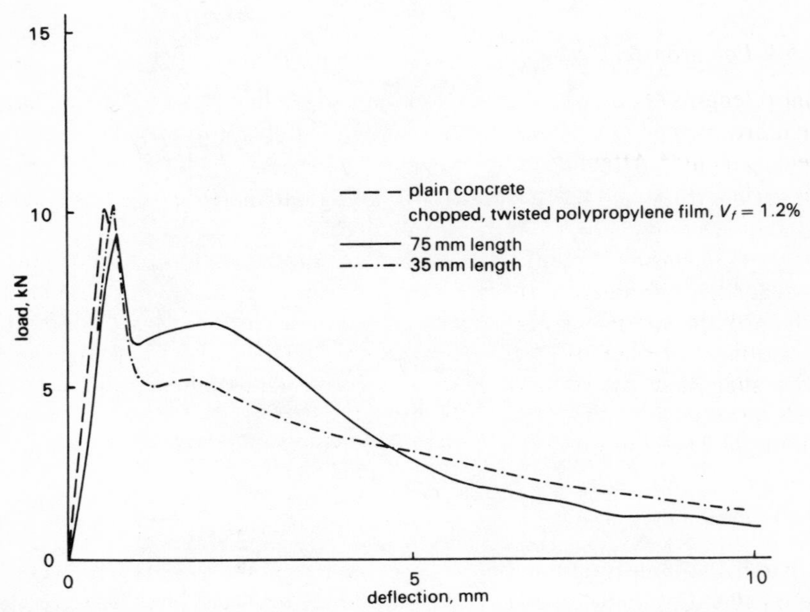

Figure 2.28 Typical load-deflection curves for a polypropylene-fibre reinforced concrete[50]

Table 2.7 Energy absorbed by fibre concretes in slow flexure[51] $V_f = 1.2\%$; age at test 2 months.

Fibre type	*Fibre size*	*Energy absorbed at a deflection of 10 mm (Nm)*
Fibrillated polypropylene	12000 denier × 35 mm long	30 (11%, 6)
Fibrillated polypropylene	12000 denier × 75 mm long	43 (32%, 6)
Stainless steel melt extract	0.7 mm dia × 25 mm long	47 (12%, 6)
Steel, Duoform type	0.38 × 38 mm	30 (19%, 6)
Steel, crimped	0.5 × 50 mm	66 (13%, 6)
Steel, hooked ends	0.5 × 50 mm	85 (27%, 5)
Plain concrete		2.8 (18%, 11)*

*Failure of plain concrete occurred before 5 mm deflection reached.
(Figures in parenthesis are the coefficient of variation followed by the number of test results on which this is based.)

specimens reinforced with steel fibres are given for comparison[51]. It is interesting to note that, while the chopped polypropylene fibre was not as effective as the best steel fibres, the 75 mm polypropylene fibre did produce specimens with an energy absorption comparable to that for the less efficient of the steel fibres, and at a considerably lower cost. Similar trends emerge if the impact strength is assessed by use of a pendulum impacting machine.

2.5.4 *Polypropylene-reinforced cement or mortar*

Short lengths of polypropylene monofilaments or film fibres have been used to reinforce cement or mortar with a view to application as asbestos-cement replacement[52]. Attempts have been made to improve the bond with the matrix by surface treatments and fraying the fibre edges to increase the surface contact and interaction between matrix and fibre[53]. An alternative approach to improve anchorage in the matrix has been to produce button-headed fibres[54].

A further approach to the bonding problem is to adopt continuous fibres whereby the composite must fail by fibre fracture rather than fibre pull-out. The utilization of many layers of opened fibrillated networks of polypropylene was suggested by Hannant and Zonsveld, with sufficient fibre volume incorporated such that composite strengths may be well in excess of matrix strength. The composite has been called 'Netcem'[55,56].

2.5.5 *Production of polypropylene reinforced cement*

Up to 8% volume fraction of short fibres have been satisfactorily incorporated into cement paste or mortars using laboratory mixers. The matrix has either had a high initial water: cement ratio (0.5–1.0) which is subsequently reduced by suction and pressing to a value of about 0.4[52], or the fresh mix is rendered liquid-like by the use of a superplasticizing agent[54].

In the laboratory, opened fibrillated networks of polypropylene film may be penetrated by hand working a superplasticized mortar matrix into successive layers of film. Fibre volumes up to 20% may be incorporated in this manner. A commercial production process to produce thin cement sheets will comprise the following steps: mixing of the cement slurry; reeling off layers of opened fibrillated film; combining the two on a suitable substrate with or without suction applied to remove excess water; forming the sheet flat or corrugated and finally curing the end product at ambient or elevated temperature. While many steps are similar to those in asbestos-cement production, important differences in the preparation and laying up of the networks have made it necessary to design new plant. For factory-scale production, network packs can now be produced with a proportion of film laid transversely for two-dimensional strength. Polypropylene networks available in this fashion have been given the trade name 'Retiflex'[57]. Typically, five packs of 1.2 m width, each containing about 12 individual layers of opened networks, can be fed from reels for continuous impregnation with cement slurry to produce 6-mm thick sheets with required 2-D strength.

Encouraging results have also been reported by Gardiner and Currie[58] on the use of woven polypropylene fabric in cement composites. The great advantage of using woven fabrics is the easy handling of the reinforcement and the ease with which they can be incorporated in cement matrices. Preliminary tests show that such composites can give flexural strength properties comparable to those of composites made with opened networks of fibrillated fibres.

2.5.6 *Properties of polypropylene-reinforced cement*

Direct tensile behaviour. This section refers to cement mortar reinforced by continuous networks of polypropylene, since tensile data for short-fibre reinforcement have not been reported.

Provided the critical fibre volume is exceeded (about 3–4% volume of aligned fibres) then the composite exhibits a tensile stress–strain curve typified by Fig. 2.6(*c*) with multiple cracking of the matrix followed by further stretching of the film reinforcement prior to fibre fracture. Figure 2.29 shows tensile stress–strain curves for a composite containing 3–7% by volume of aligned fibres of initial modulus about 8 GN/m^2. The effect of increasing the fibre volume content, and/or increasing the fibre modulus, is to reduce the increase in strain due to multiple cracking and to increase the stiffness of the final rising part of the curve. The matrix failure strain may also be increased (see Fig. 2.4) due to the inclusion of fibres; hence the cracking stress may rise since the uncracked modulus will remain effectively unaltered.

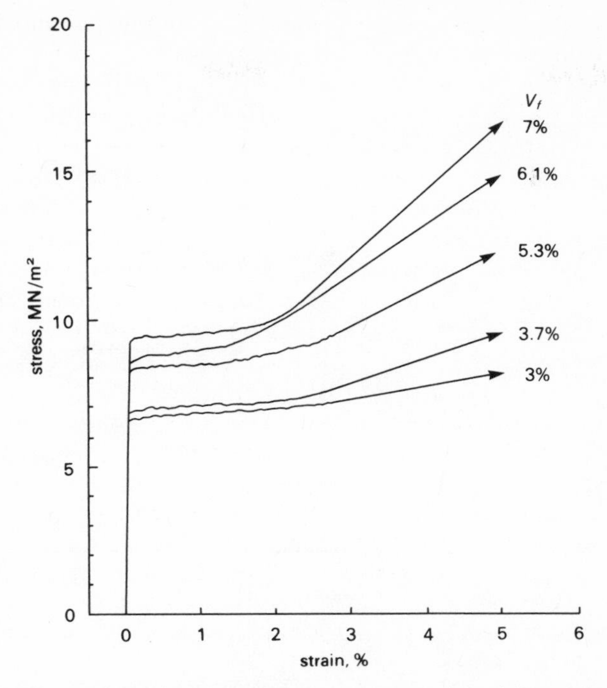

Figure 2.29 Tensile stress–strain curves for cement composite reinforced by networks of fibrillated polypropylene film

It is desirable in a thin cement sheet that cracks, should they form, are as fine as possible. This implies that the final crack spacing in tension when multiple cracking is complete should be small, e.g. if crack widths in a composite under a 2% tensile strain are to be less than 10 microns, crack spacing should not exceed 0.5 mm. Theoretical principles (equation 2.4) suggest that crack spacing is minimized by having a high volume of fibre with a high interfacial bond strength. The former is unattractive economically, so work has concentrated upon optimizing film properties to yield high bond strengths. This has highlighted the importance of fibre–matrix misfit, e.g. the interaction between matrix and fibre caused by non-uniform fibre profile and fibre surface asperities[59]. The integrity of a finely-cracked composite may, moreover, be restored by a process of autogenous healing[60].

Flexural behaviour. Load–deflection curves for specimens reinforced by various volume fractions of 26-mm long, 49-micron diameter filaments of polypropylene are shown in Fig. 2.30[52]. The ultimate flexural strengths of the higher fibre volume specimens are greater than the strengths at initial cracking, although there is a drop in load after initial cracking.

A typical rising load–deflection curve for a composite containing 6% by

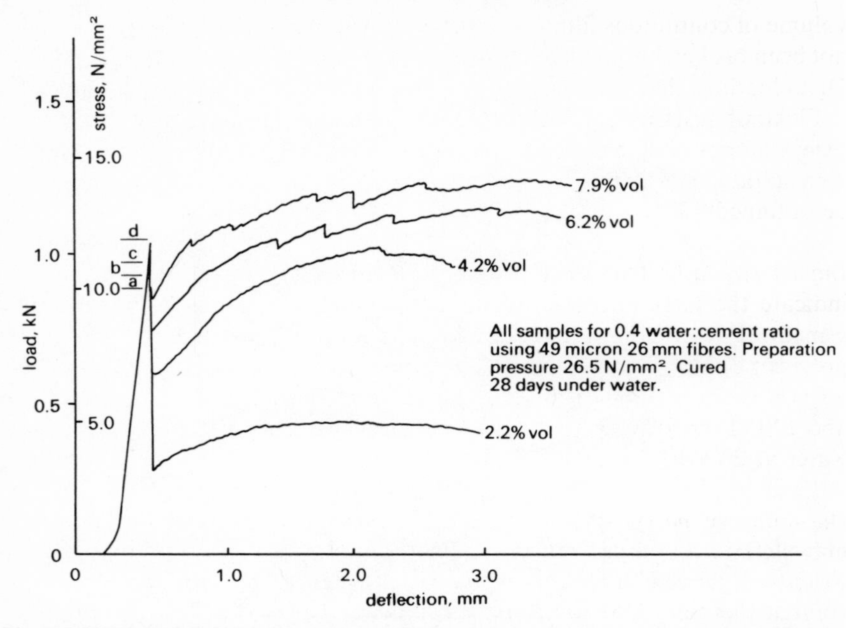

Figure 2.30 Typical flexural stress versus deflection curves for cement composite reinforced by 26 mm long filaments of polypropylene[52]

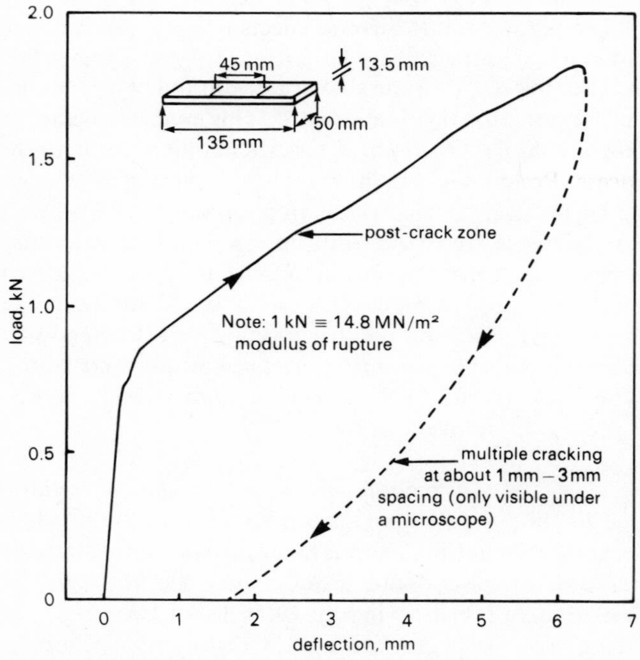

Figure 2.31 Load–deflection curve for a cement composite containing 6% by volume of opened polypropylene film networks[56]

volume of continuous film networks is shown in Fig. 2.31. The specimen has not been broken but loaded to an equivalent MOR value of about 27 MN/m^2. On unloading, the deflection returns to about 30% of the deflection under load.

Flexural response of cement composites incorporating woven polypropylene fabrics of different geometry is also encouraging, and results show that with suitable fabrics flexural performance comparable to Netcem and grc can be obtained[61].

Impact strength. The areas beneath the curves of Figs. 2.29, 2.30 and 2.31 indicate the large energy-absorbing capability of polypropylene-reinforced cement. For a composite containing about 6% by volume of aligned polypropylene fibres, the energy absorbed in failing the material in direct tension is about 1000 kJ/m^3, compared to 5.5 kJ/m^3 for an asbestos-cement and 120 kJ/m^3 for grc at 28 days, or 3 kJ/m^3 for grc after 5 years' storage in water at 20°C[62].

Durability of polypropylene-reinforced cement and concrete. Polypropylene may deteriorate under attack from ultra-violet radiation or by a thermal oxidation process. The cement matrix appears to prevent the former. To combat thermal oxidation, sophisticated stabilizers have been developed by the polymer industry and incorporated into the polymer so that degradation is delayed and a good service life can be expected.

A real-time durability programme for the Netcem material has been under way for nearly five years, with no adverse effects reported at three years[63] on uncracked and cracked specimens under natural weathering and laboratory air conditions. The most appropriate accelerated durability test is, in principle, that established for polypropylene film itself [64]. This involves ageing composite specimens at elevated temperatures in an air-circulating oven, to accelerate the oxidation process. Preliminary data from tests suggest a satisfactory long-term performance for the material, although attention has been drawn to the problem which may arise should an autoclaving process at elevated temperature followed by oven drying be adopted[65].

2.5.7 *Applications of polypropylene-reinforced cement and concrete*

It was mentioned in section 2.1.2 that the early development work on short polypropylene fibres in concrete was carried out by Shell International Chemical Co. Ltd., who called the material Caricrete[66]. Applications have included the following.

(i) Pile shells—more than half a million units have been made annually, (by West Piling and Construction Co. Ltd., U.K.) containing about 0.4% by volume of 40 mm lengths of fibrillated twine.
(ii) Cladding panels—in one example, incorporation of polypropylene

Figure 2.32 Energy-dissipating blocks of polypropylene-fibre concrete used in harbour wall construction

fibres instead of steel mesh reinforcement allowed a reduction in panel thickness from 63 to 33 mm.

(iii) Flotation units for marinas—Walcon Ltd. in the U.K. developed buoyancy units to carry jetties and walkways. A core of expanded polystyrene is surrounded by a casing of 18 mm thick polypropylene concrete, making use of the impact resistance of the material and the resistance to corrosion even after damage has occurred.

(iv) Two tonne energy-dissipating blocks in harbour wall construction (Fig. 2.32).

(v) Shotcreting—mentioned in section 2.5.2. In one application a river wall was repaired and the polypropylene fibres eliminated the need for steel (corrodable) mesh on which to spray the mortar.

The use of 0.1–0.2% by volume of chopped polypropylene filaments to stabilise an aerated concrete mix in the production of cladding has been discussed in section 2.5.2. The material was called 'Faircrete', a shortened version of fibre-air-concrete.

The potential market for polypropylene reinforced cement is, principally, as a substitute for asbestos-cement roofing and cladding panels[57]. Already polypropylene networks have been included in asbestos-cement sheets to provide increased resistance to impact loading. A typical load–deflection curve for a corrugated polypropylene-reinforced sheet compared with an asbestos-cement sheet of similar profile, is shown in Fig. 2.33[67]. Both sheets were tested over a span of 1.38 m and the load is given as a uniformly distributed load per m^2 of plan area between supports. The laboratory-made

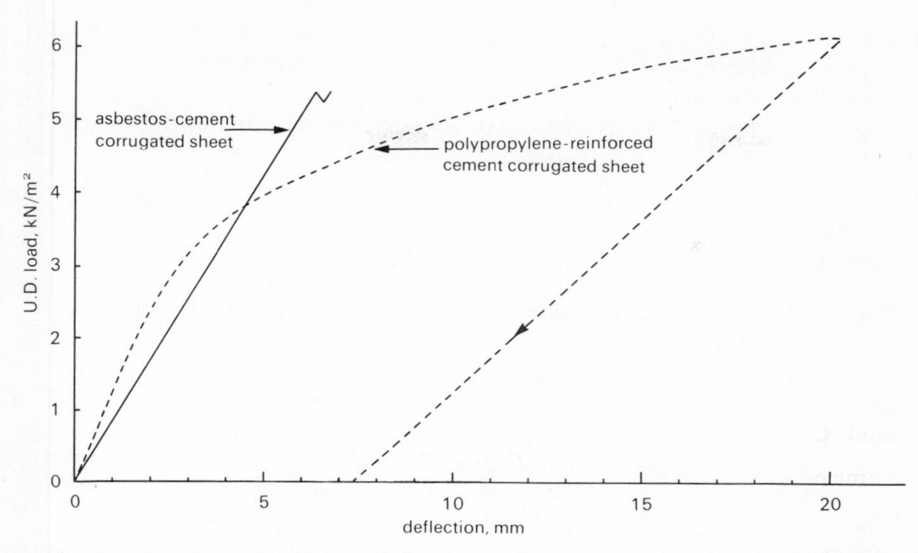

Figure 2.33 Equivalent uniformly distributed load against deflection for similar polypropylene-reinforced and asbestos-cement corrugated sheets[67]

sheet was reinforced with 8% by volume of polypropylene networks, 5% in the main longitudinal direction and 3% transversely. Under normal design loads, polypropylene-reinforced sheet is uncracked, with the post-cracking behaviour providing a considerable margin of safety against collapse and a considerable impact resistance compared to the asbestos-cement sheet.

A possible application for polypropylene-reinforced cement, which takes advantage of the high strain capacity and associated fine cracking, is as combined permanent shuttering and surface reinforcement to simplify construction and improve the durability of concrete bridge decks[68].

Gardiner *et al.*[69] have reported the manufacture and testing of several civil engineering products made in the laboratory and incorporating woven polypropylene fabrics in a cement matrix. The results are very encouraging.

2.6 Other fibre reinforcements

The fibres discussed so far have received the greatest attention in the literature and account for the bulk of the applications from the field of man-made, non-asbestos fibres. Polypropylene-reinforced cement has yet to prove itself commercially as a roofing and cladding material, but it has been the subject of considerable development work to this end.

Other fibres for cement or concrete reinforcement fall into three categories. In the first, laboratory studies have been reported in the literature but commercial development has not yet reached any advanced stage, principally due to the higher cost of the fibres and their small-scale production. Carbon,

Kevlar and high modulus polyethylene fibres fall into this classification. In the second category, asbestos-cement replacement products are or will be emerging with, in some cases, little published data in the scientific literature about the composite behaviour, e.g. cement reinforced with polyvinyl alcohol (PVA) fibres. The third category covers natural fibres and is the subject of another chapter in this book.

2.6.1 *Carbon, Kevlar and polyethylene fibre reinforcements*

Some properties of carbon, Kevlar and high-modulus polyethylene fibres have been included in Table 2.1.

The high strength and stiffness of *carbon* fibres can be used to increase the uncracked stiffness, cracking stress or strain (Fig. 2.3) and the ultimate strength of cement paste or mortar. Tensile stress–strain curves for cement reinforced by continuous, aligned carbon fibres are shown in Figure 2.34[14]. The dashed lines are the corresponding theoretical relationships obtained from the theory in section 2.2.3. The composites were manufactured by a hand lay-up technique.

Kevlar is the proprietary name for an aromatic amide polymer marketed by Dupont De Nemours International SA. A modified spray-suction technique has been used for composite fabrication, in which an atomized slurry of cement and an air stream containing 51 mm long fibres were directed simultaneously into the flat surface of a mould to a depth of 10 mm. Typical tensile stress–

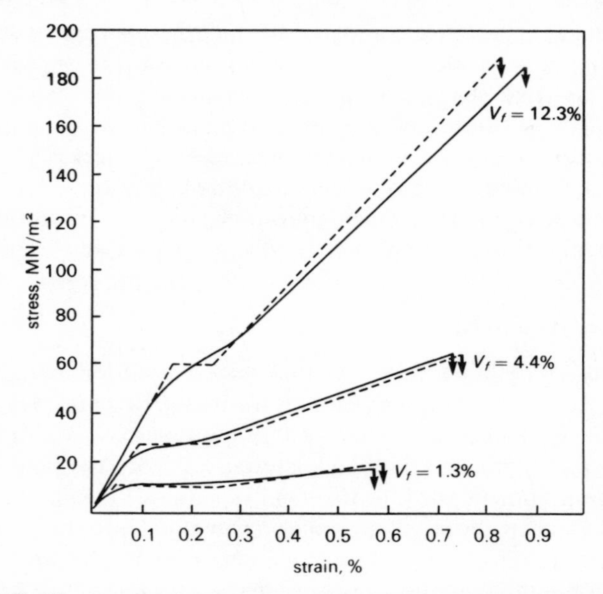

Figure 2.34 Tensile stress–strain curves for continuous carbon-fibre reinforced cement[14]:——— experimental------theoretical

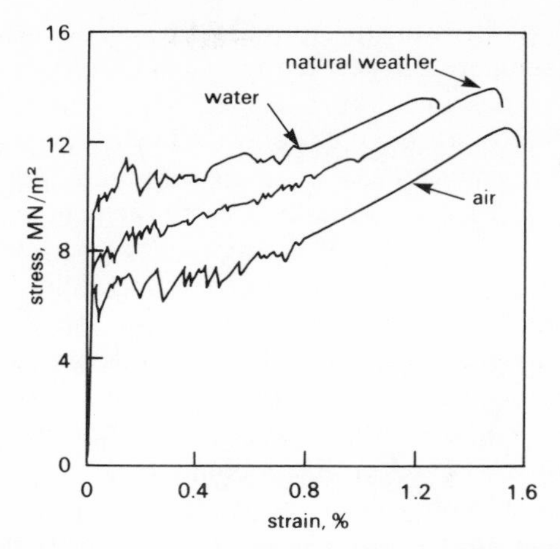

Figure 2.35 Typical tensile stress–strain relationships for a cement composite containing 2% by volume of Kevlar fibres[73]

strain relationships are shown in Fig. 2.35 for a composite containing about 2% by volume of fibres[70]. The material exhibits the familiar multiple cracking region followed by a region up to failure associated with the stretching of the fibre alone. It has been reported that one of the constituents used in the production of Kevlar fibres is carcinogenic to rats[70]. It must be confirmed that there is no health hazard before the use of Kevlar in building materials can be recommended.

Recent developments in the production of *high-modulus polyethylene* are of interest for the reinforcement of cement. For example, a superdrawn tape, Grade X40, of polyethylene has been produced by the Metal Box Company in the U.K., based on work by Ward[71]. The tape has a secant modulus of 34 GN/m^2, tensile strength of 500 MN/m^2 and an elongation at break of 5%. Polyethylene fibres with moduli in excess of 100 MN/m^2 have been produced in laboratory conditions. The drawn polyethylene fibrillates in a similar way to drawn polypropylene and has been used to produce laboratory specimens of a thin cement sheet reinforced by layers of opened networks[72].

2.6.2 *Fibres for asbestos-cement replacement*

Natural *cellulose* fibres have the advantage that they are cheap and the disadvantage that they are hygroscopic. The composite strength is reduced by the absorption of water. The dimensions of the fibre may not be stable under varying moisture content and the fibres could rot if kept moist for long periods. However, in Scandinavia, cellulose substitution for asbestos was carried out during the two world wars and inspection of such products has shown that, under certain conditions with proper treatment, the cellulose fibres would withstand outdoor exposure over 30–40 years. More recently, cellulose-reinforced cement sheets were produced in Scandinavia using Hatschek asbestos-cement machinery, but marketing awaits the results of long-term full-scale tests, which remain uncertain[73].

From the early 1970s, a range of autoclaved cellulose-reinforced calcium silicate boards has been developed containing between 2 and 10% by weight of fibre and manufactured using essentially asbestos-cement machinery[74]. Prior to development, a large number of alternative fibres were examined with reference to dispersability in water and film-forming ability, alkali resistance, temperature resistance, reinforcement effectiveness, toughness contribution and cost. The only fibres found to offer sensible levels of reinforcement, in the context of the product targets and production process used, were cellulose, carbon and Kevlar. Cellulose fibres were clearly cheaper and were eventually used to produce a variety of boards. A combination of compressing the formed corrugated shape to reduce its thickness and subsequent autoclaving has produced corrugated sheets suitable for external purposes[75]. A similar development has been reported in Australia[76]. A process of refining the fibres results in fibrillation, which enables the fibres to prevent rapid drainage and loss of matrix particles during the Hatschek production process.

Typical direct tensile stress–strain curves for a cellulose-reinforced calcium silicate matrix are shown in Fig. 2.36(*a*) for wet and dry specimens. The lower strength of the wet material is believed to be due to reductions in the modulus of the fibre and in the interfacial bond when wet, rather than to a loss of strength of fibres. The load–deflection curves of similar cellulose and asbestos-reinforced calcium silicate composite specimens are shown in Fig. 2.36(*b*), in which the greater energy-absorbing capability of the cellulose-reinforced material is apparent[74].

Other recent developments in the search for an asbestos-cement replacement have reinforced cement matrices, containing cellulose as a filler, with short acrylonitrile fibres[77] cut from a fibrillated film or with short polyvinyl alcohol (PVA) fibres[78]. Presumably in these cases the cellulose has a secondary role as reinforcement, its primary function being to prevent segregation of the cement during the Hatschek process of manufacture.

2.6.3 *Combinations of fibres*

Fibre-reinforced cements and concretes have been produced incorporating a mixture of fibres, in attempts to optimize the benefits derived from the individual fibres. Walton and Majumdar[79] have considered combinations of low-modulus, short polymer fibres—polypropylene and nylon—with high-modulus fibres such as asbestos, glass and carbon. The principal benefit of the combination is the higher impact strength imparted by the low modulus fibres.

According to Kobayashi and Cho[80], it is possible to obtain a fibre-reinforced concrete of superior toughness by dispersing short steel and

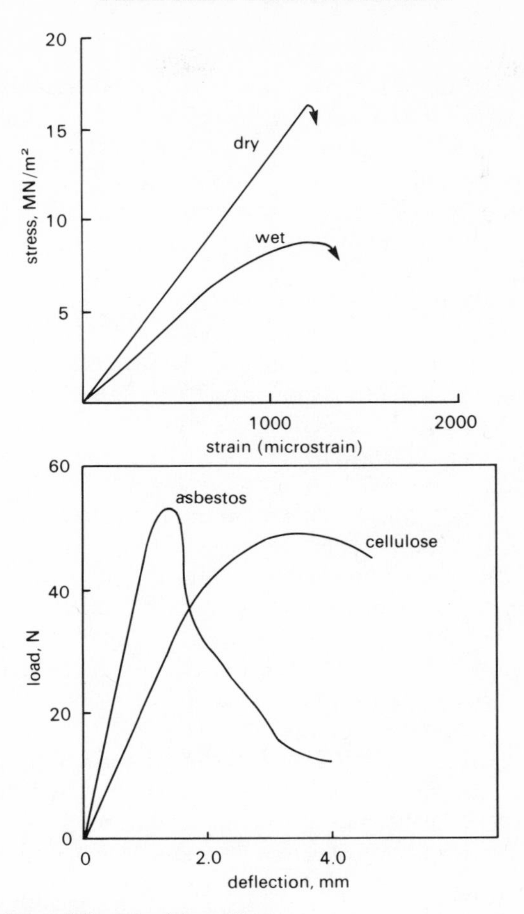

Figure 2.36 Cellulose-reinforced calcium silicate:
(a) tensile stress–strain curves, wet and dry specimens[73]
(b) load–deflection curve compared to a similar asbestos-reinforced calcium silicate specimen[74]

polyethylene fibres in randomly orientated states in concrete. Within a practical range, the optimum combination was obtained with 1% by volume of steel fibres and 3% by volume of polyethylene fibres.

Hughes[81] has combined continuous rovings of Kevlar with grc in order to overcome the limited tensile strength and increasing brittleness with age of plain grc. The combination is referred to as AGRC, and has been considered for the production of lightweight structural units such as thin-walled trough sections. The Kevlar roving performs the same function as conventional bar reinforcement and the grc provides shear resistance and improved impact resistance compared to the unreinforced matrix, despite the fact that the impact strength is reduced with ageing.

2.7 Concluding remarks

There is considerable incentive to maximize the use of cheap, low-energy-cost materials such as mortar and concrete. The object of fibre reinforcement is to control the brittle behaviour of cement-based materials and so increase their attractiveness, to the construction industry particularly.

The use of steel-fibre reinforced concrete is now established practice as an economic solution to a number of engineering problems. Its usage will continue to increase as the body of expertise and experience in the in-service performance spreads. In Japan, for example, 2500 tons of carbon-steel fibres and 500 tons of stainless steel fibres were used in 1981, the major application being in the lining of underground structures. Successful applications have taken particular advantage of the enhanced toughness of concrete reinforced by fibres and of the facility of including the reinforcement integrally with the bulk concrete. It is surprising that the cheaper polymer fibres, successful in certain applications, have not been more widely used.

In structural concrete, conventional reinforcing by steel bars or prestressing strands is efficient for static strength purposes. The greatest potential for steel fibres in this area is in combination with conventional reinforcement, rather than as a replacement, to control cracking, reduce local damage and improve performance under shock loadings.

In thin sheet applications, where conventional reinforcement is unsuitable, glass-reinforced cement has rapidly established itself as a construction material with its own design philosophy. The cost of glass fibres and the problem of ensuring satisfactory long term performance have mitigated against an even wider useage. Fibre and matrix developments have resulted in better durability and further improvements should be possible.

Ideally, the objective in thin sheet applications is a strong, durable, quasi-ductile cement composite. The effect of fibres should be to enhance the strength of the matrix, which once exceeded, should result in virtually invisible cracks. Considerable advances have been made towards meeting this objective, at reasonable cost, with the use of polymer fibres. Future developments in polymer fibre technology will further improve composite performance.

References

1. Austrian Patent No. 5970, applied for 30th March, 1900, accepted 15th June, 1901, Ludwig Hatschek, Vocklabruck.
2. Hayden, R. (1942) *Fundamental Problems in Asbestos Cement Production*. Building Research Station Library Communication No. 598. (Translated from the German).
3. *Study on the Opportunities of Manufacturing Asbestos Products in Quebec*. Report Phase I, 1977, submitted to Quebec Asbestos Mining Association by Sores Inc., Montreal and Arthur D. Little, Inc., Cambridge.
4. Romualdi, J.P., and Batson, G.B. (1963) Mechanics of crack arrest in concrete. *Proc. Amer. Soc. of Civ. Engrs.* **89**, 147–168.
5. Krenchel, H. (1964) *Fibre Reinforcement*. Akademisk Forlag, Copenhagen.

6. Biryukovich, K.L., Biryukovich, Y.L., and Biryukovich, D.L. (1964) *Glass Fibre-Reinforced Cement*. Published by Budivel' nik, Kiev, CERA translation.
7. Goldfein, S. (1965) Fibrous reinforcement for Portland cement. *Modern Plastics*, 156–159.
8. Swamy, R.N., Mangat, P.S., and Kameswara Rao, C.V.S. (1974) 'The mechanics of fibre reinforcement of cement matrices'. Publication SP-44, *Fibre Reinforced Concrete*, American Concrete Institute, 1–28.
9. Swamy, R.N. (1975) Fibre reinforcement of cement and concrete. *RILEM Mater. Struct.* **8** (45), 235–254.
10. Hannant, D.J. (1978) *Fibre Cements and Fibre Concretes*. John Wiley and Sons, New York.
11. Edgington, J., Hannant, D.J., Williams, R.I.T. (1974) *Steel fibre reinforced concrete*. Building Research Estab. Current Paper 69/74.
12. Kelly, A. (1974) 'Some scientific points concerning the mechanics of fiburous composites', *Composites–Standards, Testing and Design*, NPL Conf. Proc., 9–16.
13. Aveston, J., Cooper, G.A., and Kelly, A. (1971) 'Single and multiple fracture', *The properties of fibre composites*, NPL Conf. Proc. IPC Science and Technology Press, 15–24.
14. Aveston, J., Mercer, R.A., and Sillwood, J.M. (1974) 'Fibre-reinforced cements—scientific foundations for specifications', *Composites–Standards, testing and design*, NPL Conf. Proc., 93–103.
15. Korczynskyj, Y., Harris, S.J., Morley, J.G. (1981) 'The influence of reinforcing fibres on the growth of cracks in brittle martix composites', *J. Mater. Sci.* **16**, 1533–1547.
16. Hannant, D.J., Hughes, D.C., and Kelly, A. (1983) 'Toughening of cement and of other brittle solids with fibres', *Phil. Trans. Roy. Soc.* **A310**, 175–190.
17. Hannant, D.J. (1975) 'The effect of post-cracking ductility on the flexural strength of fibre cement and fibre concrete', *Proc. RILEM Symp. 1975*, Construction Press Ltd., 499–508.
18. Edgington, J. (1977) 'Economic fibrous concrete', *Fibre-Reinforced Materials*, Conf. Proc. Instn. of Civ. Engnrs., 129–140.
19. Swamy, R.N. (1974) The technology of steel fibre reinforced concrete for practical applications. *Proc. Inst. Civ. Engrs.* **56**, 143–159.
20. RILEM Technical Committee Report (1977) Fibre concrete materials. *RILEM Mater. Struct.* **10** (56), 103–120.
21. Swamy, R.N., and Stavrides, H. (1975) 'Some properties of high workability steel-fibre concrete', *Proc. RILEM Symp.* 1975, construction Press Ltd., 197–208.
22. Johnston, C.D. (1982) Steel fibre-reinforced concrete—present and future in engineering construction. *Composites* **13**, 113–121.
23. ACI Comm. 544 (1978) Measurement of properties of fibre reinforced concrete. *ACI Journal* **75**, 283–289.
24. Swamy, R.N., and Jojagha, A.H. (1982) Workability of steel fibre reinforced lightweight aggregate concrete. *Int. J. Cem. Comp. and Lightweight Conc.* **4 (2).** 103–109.
25. Shah, S.P., Stroeven, P., Dalhuisen, D., van Stekelburg, P. (1978) 'Complete stress–strain curves for steel fibre reinforced concrete in uniaxial tension and compression', *Proc. RILEM Symp. 1978*, Construction Press Ltd., 399–408.
26. Swamy, R.N., and Al-Noori, K.A. (1975) 'Flexural behaviour of fibre concrete with conventional steel reinforcement', RILEM Symposium 1975, *Fibre Reinforced Cement and Concrete*, The Construction Press Ltd., Hornby, 187–196.
27. Johnston, C.D. (1980) 'Properties of steel fibre reinforced mortar and concrete', *Proc. Symp. on Fibrous Concrete*, Construction Press Ltd., 29–47.
28. Johnston, C.D. (1974) *Steel fibre-reinforced mortar and concrete: a review of mechanical properties*. ACI Publ. SP-44, 127–142.
29. Swamy, R.N. (1980) Influence of slow crack growth on the fracture resistance of fibre cement composites. *Int. J. Cem. Comp.* **2** (1), 43–53.
30. ACI Committee 544 (1982) State-of-the-art report on fibre-reinforced concrete. Report no. ACI 544 IR-82, *Concrete International, Design and Construction*, **4** (5), 9–30.
31. Swamy, R.N., and Jojagha, A.H. (1982) Impact resistance of steel fibre reinforced lightweight aggregate concrete. *Int. J. Cem. Comp. and Lightweight Concrete* **4** (4), 209–220.
32. Swamy, R.N., and Lankard, D.R. (1974) Some practical applications of steel fibre reinforced concrete. *Proc. Instn. of Civ. Engrs.* **56**, 235–256.
33. Nishioka, K., Yamakawa, S., Kameda, Y., and Akihama, S. (1980) Present status of applications of steel fibre concrete in Japan. *Int. J. Cem. Comp.* **2** (4), 205–232.

34. Kobayashi, K. (1983) Development of fibre reinforced concrete in Japan. *Int. J. Cem. Comp. and Lightweight Concrete* **5** (1), 27–40.
35. Williams, R.I.T. 'Steel fibre concrete in road and airfield pavements', Chapter 12 in reference 10, 182–197.
36. Josifek, C.W. (1980) 'Fibre concrete aircraft parking ramp at McCarran International Airport'. *Int. J. Cem. Composites* **2**, 235–238.
37. Swamy, R.N. and Kent, B. (1974) Some practical structural applications of steel fibre reinforced concrete. Publication SP-44, *Fibre Reinforced Concrete*, American Concrete Institute, 319–336.
38. Swamy, R.N. (1980) 'Prospects of fibre reinforcement in structural applications', *Proc, Symp. on Advances in Cement-Matrix Composites*, Materials Research Society, Boston 159–169.
39. Fordyce, M.W., and Wodehouse, R.G. (1983) *GRC and Buildings*. Butterworth and Co. Ltd., London.
40. Ali, M.A. Majumdar, A.J., and Singh, B. (1974) *Properties of glass fibre cement—the effect of fibre length and content*. Building Research Estab. CP 79/74.
41. Majumdar, A.J. (1980) 'Properties of grc', *Proc. Symp. on Fibrous Concrete*, Construction Press Ltd., 48–68.
42. Majumdar, A.J. (1980) 'Some aspects of glass fibre reinforcement cement research', *Proc. Symp. on Advances in Cement-Matrix Composites*, Materials Research Society, Boston 37–59.
43. Proctor, B.A. (1980) 'Properties and Performance of GRC', *Proc. Symp. on Fibrous Concrete*, Construction Press Ltd., 69–86.
44. Proctor, B.A. (1981) 'Past Developments and Future Prospects for GRC materials', *Proc. Int. Congr. on Glass Fibre Reinforced Cement*, Glassfibre Reinforced Cement Association, 50–67.
45. Jacobs, M.J.N. (1981) 'Forton PGRC—a many-sided construction material', *Proc. Int. Congress on Glass Fibre Reinforced Cement*, Glassfibre Reinforced Cement Assoc, 31–49.
46. Bijen, J. (1983) Durability of some glass fibre reinforced cement composites. *ACI Journal*, Jul-Aug 1983, 305–311.
47. Smith, J.W. (1983) The replacement of asbestos cement by GRC. *Composites* **13**, 161–163.
48. Ritchie, A.G.B., and Al-Kayyali, O.A. (1975) 'The effects of fibre reinforcements on lightweight aggregate concrete', *Proc. RILEM Symp.* 1975, Construction Press Ltd, 247–256.
49. Hobbs, C. (1971) 'Faircrete: an application of fibrous concrete', *Proc. of Int. Building Exhibition Conf*., Publ. Building Research Estab. 59–67.
50. Hibbert, A.P. (1979) Ph.D. Thesis, University of Surrey.
51. Hibbert, A.P. and Hannant, D.J. (1981) *Impact resistance of fibre concrete*. Transport and Road Research Laboratory Supplementary Report 654.
52. Dave, N.J., and Ellis, D.G. (1979) Polypropylene fibre reinforced cement. *Int. J. Cem. Composite* **1**, 19–28.
53. Krenchel, H., and Jensen, H.W. (1980) 'Organic reinforcing fibres for cement and concrete', *Proc. Symp. on Fibrous Concrete*, Construction Press Ltd, 87–98.
54. Naaman, A.E., Shah, S.P., and Throne, J.L., Some developments in polypropylene fibres for concrete. *ACI Journal*, to be published.
55. Hannant, D.J., and Zonsveld, J.J. (1980) Polyolefin fibrous networks in cement matrices for low cost sheeting. *Phil. Trans. Roy. Soc.* **A294**, 591–597.
56. Hannant, D.J., Zonsveld, J.J., and Hughes, D.C. (1978) Polypropylene film in cement-based materials. *Composites* **9**, 83–88.
57. Vittone, A. (1983) 'Netlike products in polypropylene fibrillated films—production and applications', *3rd Int. Conf. on Polypropylene Fibres and Textiles*, Plastics and Rubber Institute, 40.1–40.10.
58. Gardiner, T., and Currie, B. (1983) Flexural behaviour of composite cement sheets using woven polypropylene mesh fabrics. *Int. J. Cem. Comp. and Lightweight Concrete* **5** (3), 193–197.
59. Hughes, D.C. (1983), Ph.D. Thesis, University of Surrey.
60. Hannant, D.J., and Keer, J.G. (1983) Autogenous healing on thin cement-based sheets. *Cem. Concr. Res.* **13**, 357–365.
61. Gardiner, T., Currie, B., and Green, H. (1983) 'Flexural behaviour of a cement matrix reinforced with polypropylene mattings', *3rd Int. Conf. on Polypropylene Fibres and Textiles*, Plastics and Rubber Institute, 38.1–38.8.
62. Hibbert, A.P., and Hannant, D.J. (1982) 'Toughness of cement composites containing polypropylene films compared with other fibre cements', *Composites* **13**, 393–399.

63. Hannant, D.J. (1983) 'Durability of cement sheets reinforced with fibrillated polypropylene networks', *Mag. Concr. Res.* **35** (125), 197–204.
64. ANSI/ASTM D3045–74 (Reapproved 1979) *Heat Ageing of Plastics Without Load.*
65. Mai, Y.W., Andonian, R., and Cotterell, B. (1980) 'Thermal degradation of polypropylene fibres in cement composites', *Int. J. Cem. Comp.* **2**, 149–155.
66. Zonsveld, J.J., (1970) 'The marriage of plastics and concrete', *Plastica* **23**, 474–484.
67. Keer, J.G. and Hughes, D.C. (1982) 'Polypropylene-reinforced cement sheeting for use as roofing and cladding elements', *Proc. of 4th Int. Conf. on Composite Materials*, ICCM-IV, Volume 2, 1255–1262.
68. Galloway, J.W., Williams, R.I.T., and Raithby, K.D. (1981) *Mechanical properties of polyolefin-reinforced cement sheet for crack control in reinforced concrete.* Transport and Road Research Supplementary Report 658.
69. Gardiner, T., Currie, B., and Green, H. (1983) 'Performance of civil engineering products made from a cement matrix reinforced with polypropylene mattings', *3rd Int. Conf. on Polypropylene Fibres and Textiles*, Plastics and Rubber Institute, 39.1–39.7.
70. Walton, P.L., and Majumdar, A.J. (1980) 'Properties of cement composites reinforced with Kevlar fibres', *J. Mater. Sci.* **13**, 1075–1083.
71. Ward, I.M. (1980) 'Ultra-high modulus polyolefins', *Phil. Trans. Roy. Soc.* **A294**, 473–482.
72. Hughes, D.C., and Hannant, D.J. (1982) 'Brittle matrices reinforced with polyalkene films of varying elastic moduli', *J. Mater. Sci.* **17**, 508–516.
73. Pedersen, N. (1980) 'Commercial development of alternative to abestos sheet products based on short fibres', *Proc. Symp. on Fibrous Concrete*, Construction Press Ltd., 189–193.
74. Harper, S. (1982) Developing asbestos-free calcium silicate building boards. *Composites* **13**, 123–128.
75. UK Patent Application GB 2105636 A, publ. 30 March 1983.
76. Coutts, R.S.P. (1983) Wood fibres in inorganic matrices. *Chemistry in Australia* **50**, 143–148.
77. U.K. Patent Application GB 2065735 A, 1981, Amrotex AG.
78. TAC Construction Materials (1983) Duracem, trade literature.
79. Walton, P.L., and Majumdar, A.J. (1975) *Cement-based composites with mixtures of different types of fibres.* Building Research Estab. Current Paper CP 80/75.
80. Kobayashi, K., and Cho, R. (1982) 'Flexural characteristics of steel fibre and polyethylene fibre hybrid-reinforced concrete', *Composites* **13**, 164–168.
81. Hughes, B.P. (1980) 'AGRC Composites for thin structural sections', *Proc. Symp. on Advances in Cement-Matrix Composites*, Materials Research Society, 187–196.

3 Concrete reinforced with natural fibres

M.A. AZIZ, P. PARAMASIVAM and S.L. LEE

Abstract

In recent years, a great deal of interest has been created worldwide on the potential applications of natural-fibre reinforced concrete*. Investigations have been carried out in many countries on various mechanical properties, physical performance and durability of concrete materials reinforced with natural fibres from coconut husk, sisal, sugarcane bagasse, jute, wood and other vegetable fibres. These investigations have shown encouraging prospects for this new distinct group of materials for potential applications in various types of constructions.

This chapter reports chronological developments, present status and future prospects of natural-fibre reinforced concretes in various engineering constructions. Topics include fibre materials and their characteristics, production technology, factors affecting properties, properties in the fresh and hardened states, physical performance and durability. It also highlights design parameters, production procedures, construction techniques and practical applications.

Natural-fibre reinforced concrete is essentially a special-purpose concrete which consists of small-diameter discontinuous, discrete natural fibres of different origin randomly distributed in a cementitious matrix. The uniform dispersal of fibres in a cementitious matrix distributes stresses and enhances resistance to cracking, impact and shock loadings, and also improves ductility for better energy absorption. This new distinct group of construction materials possesses good acoustic and thermal properties.

Natural-fibre reinforced concrete can be used in conventional applications as well as in applications where energy has to be absorbed or where impact damage is likely to be encountered.

3.1 Introduction

Natural fibres of various types are abundantly available in many parts of the world. There has been a growing interest in recent years in utilizing natural

*For brevity, the term 'concrete' is used here to include cement paste, mortar and other cement based matrices.

fibres for reinforcing concrete. Although the use in concrete of short, discrete reinforcing fibres, other than natural ones has drawn world-wide interest[1–9], the use of natural fibres in concrete is relatively recent in spite of the fact that the concept of fibre reinforcement was recognized more than fifty years ago. Several examples of fibre reinforcement exist in nature as well as in the early history of mankind. Nature has provided man with fibre-reinforced materials in the form of wood, bamboo and vegetables. The use of straw in sun-dried mud brick and walls, and horse-hair in mortar predates the use of conventional reinforced concretes.

Natural fibres are prospective reinforcing materials and their use until now has been more traditional than technical. They have long served many useful purposes but the application of materials technology for the utilization of natural fibres as the reinforcement in concrete has only taken place in comparatively recent years. In the late 1960s and early 1970s, investigations began on the possibility of using organic fibres of various origins as reinforcement in thin concrete sheets and other cement-based composites[10–92]. These investigations soon indicated the possibility of manufacturing products of natural-fibre reinforced concrete whose properties are comparable to those of asbestos-cement products. Appropriate methods of manufacturing roofing and wall sheets and other products of natural fibre concrete began rapidly in some countries in Central America, Africa, Asia, Australia and Europe. Natural-fibre reinforced cement or concrete products using fibres such as coconut coir, sisal, sugarcane bagasse, bamboo, jute, wood and vegetables have been tested so far in more than 40 countries[47]. These tests have shown encouraging results. Economics and other related factors in many developing countries where natural fibres of various origin are abundantly available, demand that construction engineers and builders apply appropriate technology to utilize these natural fibres as effectively and economically as possible to produce good-quality fibre-reinforced composite materials for housing and other needs.

Natural fibres are available in most developing countries and require only a low degree of industrialization for their processing. In comparison with an equivalent volume or weight of the most common synthetic reinforcing fibres, the energy required for their production is small and hence the cost of fabricating these composites is also low. In addition, the use of a random mixture of natural fibres in concretes leads to a technique that requires only a small number of trained personnel in the construction industry. The use of such fibres in concrete is therefore particularly attractive to the developing countries with their shortage of skilled manpower and capital, and their need for good quality, locally-produced low-cost building materials.

3.2 Fibre materials

The basic requirements of natural fibres when used as reinforcement in concrete matrices are high tensile strength and elastic modulus, reasonable bond at the interface with the matrix, good chemical and geometric stability and durability. Types of natural fibres investigated as prospective reinforcing materials include coconut fibre, sisal fibre, sugarcane bagasse fibre, bamboo fibre, palm fibre, jute fibre, flax fibre, wood fibre and some vegetable fibres. The properties of most natural fibres are fairly well established.

3.2.1 *Coconut fibre*

Coconut cultivation is concentrated in the tropical belts of Asia and East Africa. The outer covering of fibrous material of a matured coconut, termed coconut husk, is the reject of a coconut fruit. The husk of the coconut consists of hard skin and numerous fibres embedded in a soft cork-like material usually referred to as pith. The fibres are normally 150 to 350 mm long and consist mainly of lignin, tannin, cellulose, pectin and other water-soluble substances[47,93]. Fibres are usually extracted by the process known as retting in which tannins and pectins are dissolved in water and most of other substances are decomposed. Fibres are also separated from the husk by a mechanical process. The physical properties of coconut fibres vary slightly from region to region. Typical properties of a commercial variety are given in Table 3.1. Figure 3.1 shows a typical stress–strain relation of coconut fibres.

3.2.2 *Sisal fibre*

Sisal is one of the strongest natural fibres[47,94]. Its traditional use as a reinforcement for gypsum plaster sheets in the Australian building industry[26] has created interest in some research groups and construction firms especially in Sweden[40–58] during the last few years to utilize these fibres to produce

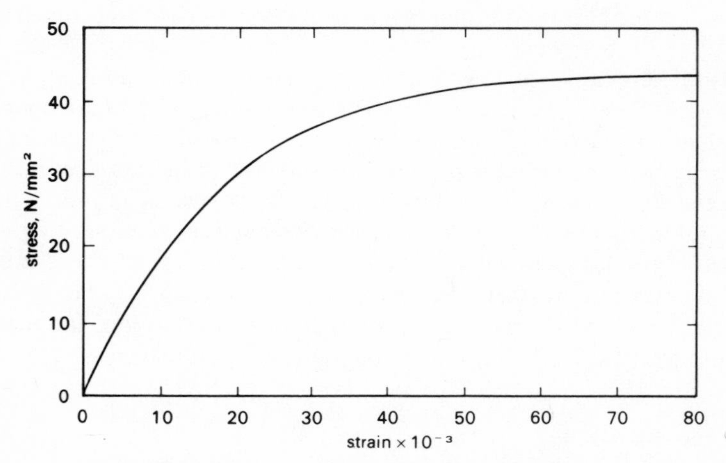

Figure 3.1 Typical stress–strain relation of coconut fibres[34,39]

Table 3.1 Typical properties of coconut fibres[34,35,39,47]

1. Specific gravity	1.12–1.15
2. Bulk density, kg/m^3	145–280
3. Fibre length, mm	50–350
4. Fibre diameter, mm	0.10–0.40
5. Ultimate tensile strength, N/mm^2	120–200
6. Modulus of elasticity, kN/mm^2	19–26
7. Elongation at break, percent	10–25
8. Water absorption, percent	130–180

good-quality natural-fibre reinforced concretes. There are a number of grades of commercial sisal fibre differing in properties. Typical properties of sisal fibres are given in Table 3.2.

3.2.3 *Sugarcane bagasse fibre*

Bagasse is the fibrous residue which is obtained in cane-sugar production after extraction of the juice from the cane stalks. Sugarcane belongs to the grass family. It grows up to 6 m high depending on species and cultivation area, and has a cane diameter of up to 6 cm. Sugarcane cultivation is concentrated in tropical and sub-tropical regions with sufficient humidity. When the bagasse leaves the sugar mill, it contains primarily fibres (50 to 55%), moisture (15 to 20%), pith (30 to 35%) and a relatively small amount of soluble solids (4 to 6%)[95]. Its composition varies according to the variety of cane, its maturity, method of harvesting, and finally, the efficiency of the milling plant. The pith is the weak constituent and is discarded during the depithing process[95]. Typical properties of bagasse fibres are given in Table 3.3 and a stress–strain relation in Fig. 3.2.

Table 3.2 Typical properties of sisal fibres[41,47,51,94]

1. Bulk density, kg/m^3	700–800
2. Ultimate tensile strength, N/mm^2	280–568
3. Modulus of elasticity, kN/mm^2	13–26
4. Elongation at break, percent	3–5
5. Water absorption, percent	60–70

Table 3.3 Typical properties of sugarcane bagasse fibres[59,60,95]

1. Specific gravity	1.20–1.30
2. Fibre length, mm	50–300
3. Fibre diameter, mm	0.2–0.4
4. Water content, percent	15–20
5. Water absorption, percent	70–75
6. Ultimate tensile strength, N/mm^2	170–290
7. Modulus of elasticity, kN/mm^2	15–19

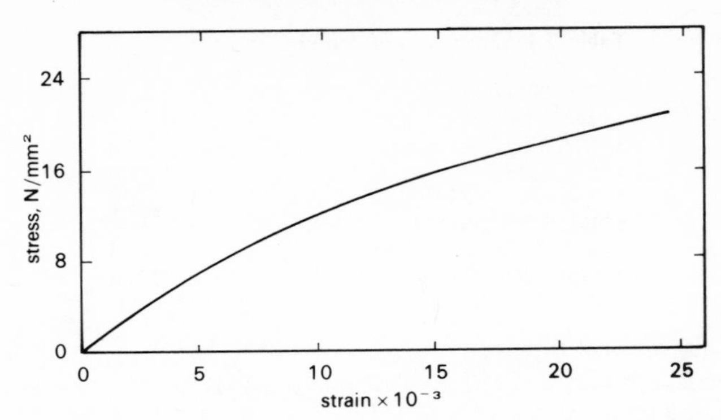

Figure 3.2 Typical stress–strain relation of sugarcane bagasse fibres[59,60]

3.2.4 *Bamboo fibre*

As a natural vegetation, bamboo is grown in abundance in the tropical and sub-tropical regions. Bamboo belongs to the grass family and can grow up to 15 m high, having diameters in the range of 25 to 100 mm. Bamboo fibres are extracted by a special device[61–65]. Bamboo fibres are remarkably strong in tension but they have low modulus of elasticity and high water absorption potential. The stress–strain relation of bamboo fibres is shown in Fig. 3.3 and typical properties are given in Table 3.4. The relationship of equivalent diameter to ultimate tensile strength and modulus of elasticity is shown in Figs. 3.4 and 3.5 respectively[65]. The values of water absorption, ultimate

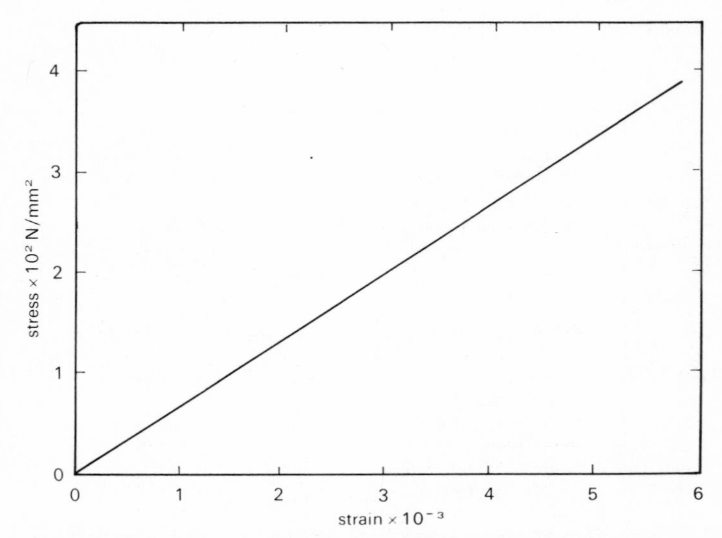

Figure 3.3 Typical stress–strain relation of bamboo fibres[65]

Table 3.4 Properties of bamboo fibres[65]

1. Specific gravity	1.52
2. Mean periphery, mm	1.24
3. Mean cross-sectional area, mm^2	0.10
4. Ultimate tensile strength	442.00
5. Modulus of elasticity, kN/mm^2	37.00
6. Bond strength (pull-out test, at 5.8% air voids), N/mm^2	1.96
7. Slip modulus, N/mm^2	77.66

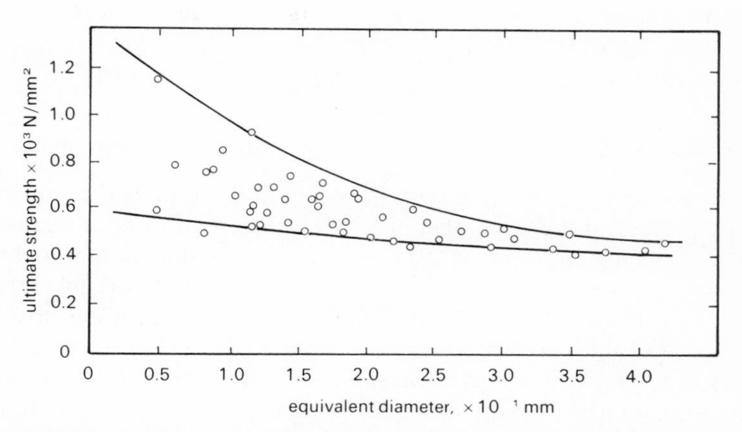

Figure 3.4 Relationship between equivalent diameter and ultimate tensile strength of bamboo fibres[65]

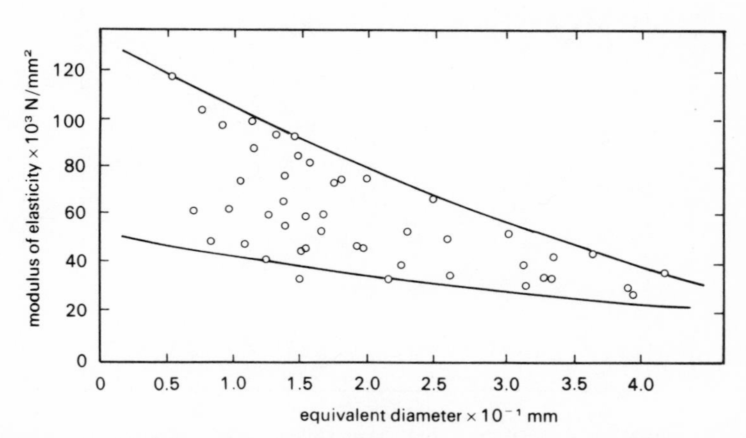

Figure 3.5 Relationship between equivalent diameter and modulus of elasticity of bamboo fibres[65]

tensile strength and modulus of elasticity are in the range of 40 to 45 percent, 350 to 500 N/mm² and 33 to 40 kN/mm² respectively[96,97].

3.2.5 *Palm fibre*

Palms are of various types and grow abundantly in many parts of the world. Fibres extracted from decomposed palm trees, especially from date-palms, are found to be brittle, having very low tensile strength and modulus of elasticity and very high water absorption[66,67,95].

3.2.6 *Jute fibre*

Jute is abundantly grown in Bangladesh, China, India and Thailand. Jute fibres are extracted from the fibrous bark of the jute plants which grow as tall as 2.5 m with a diameter of the stem at the base of around 25 mm. The method of extraction of fibres from the jute plants is simple. The matured plants are cut down, tied into bundles and submerged in water for about four weeks during which the bark is completely decomposed, exposing the fibres. The fibres are then stripped off manually from the stems, washed and sun-dried. There are different varieties of jute fibres with varying properties[68,98]. Typical stress–strain relations of jute fibre are shown in Fig. 3.6 and typical properties are given in Table 3.5.

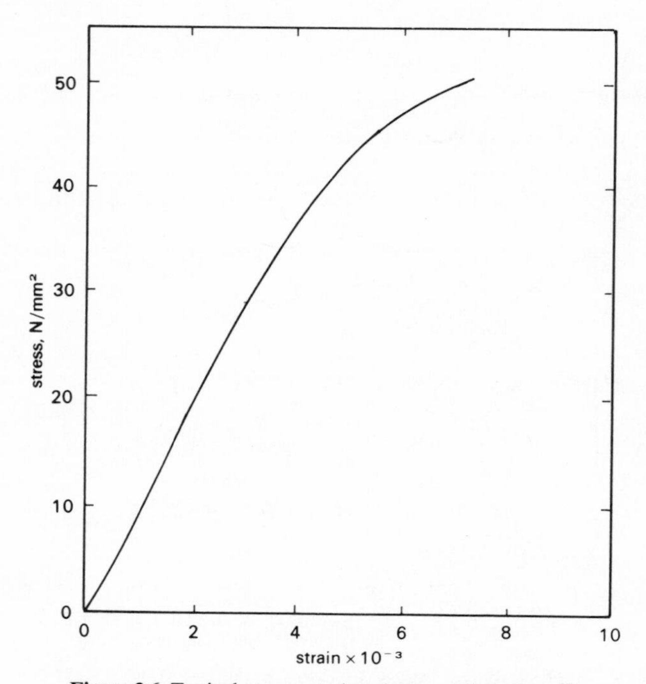

Figure 3.6 Typical stress–strain relation of jute fibres[69]

Table 3.5 Properties of jute fibres[47,68,69]

1. Fibre length, m	1.8–3.0
2. Fibre diameter, mm	0.1–0.2
3. Specific gravity	1.02–1.04
4. Ultimate tensile strength, N/mm^2	250–350
5. Modulus of elasticity, kN/mm^2	26–32
6. Elongation at break, percent	1.5–1.9

3.2.7 *Flax fibre*

Flax is a slender, erect, blue-flowered plant grown for its fibres and seeds in many parts of the world[47,87]. Flax fibres are strong and durable (Table 3.6).

3.2.8 *Wood fibre*

Wood fibres are generally obtained from softwood trees (conifers) like pine, spruce and larch. Wood chips are soaked in water containing sodium sulphate and the chips are then placed in a defibrator and presteamed for some time under a certain pressure. The defibrator is then rotated at a certain number of revolutions per minute for a definite time period[74,77]. The resulting pulp is washed and air-dried. This method produces a high-yield pulp ($>90\%$) with uncollapsed lignified fibres. Wood fibres are strong and durable[74–87].

3.2.9 *Vegetable fibres*

Different types of vegetable fibres are available in most developing countries[99,100]. Only five such fibres, namely akwara, elephant grass, water reed, plantain and musamba, have so far been found to be potential reinforcing materials[80,81].

Akwara is a vascular bundle, readily available in Nigeria and some other countries of the world, consisting of a sheath of fibres which surround an annular layer known as primary phloem, within which there are two metaxylem cavities. The phloem consists of broken-down materials, the sheath is made up of numerous fibre cells, and akwara fibres are obtained from these cells. The geometry of a fibre is variable; it may be circular, rectangular, or elliptical in cross-section, tapering along the length. The equivalent diameter varies between 1 and 4 mm. Fibres are usually 1.5 m long. Specific gravity is around 0.96. Fig. 3.7 shows the typical stress–strain behaviour of

Table 3.6 Properties of flax fibres[47,87]

1. Fibre length, m	0.5
2. Ultimate tensile strength, N/mm^2	1000
3. Modulus of elasticity, kN/mm^2	100
4. Elongation at break, percent	1.8–2.2

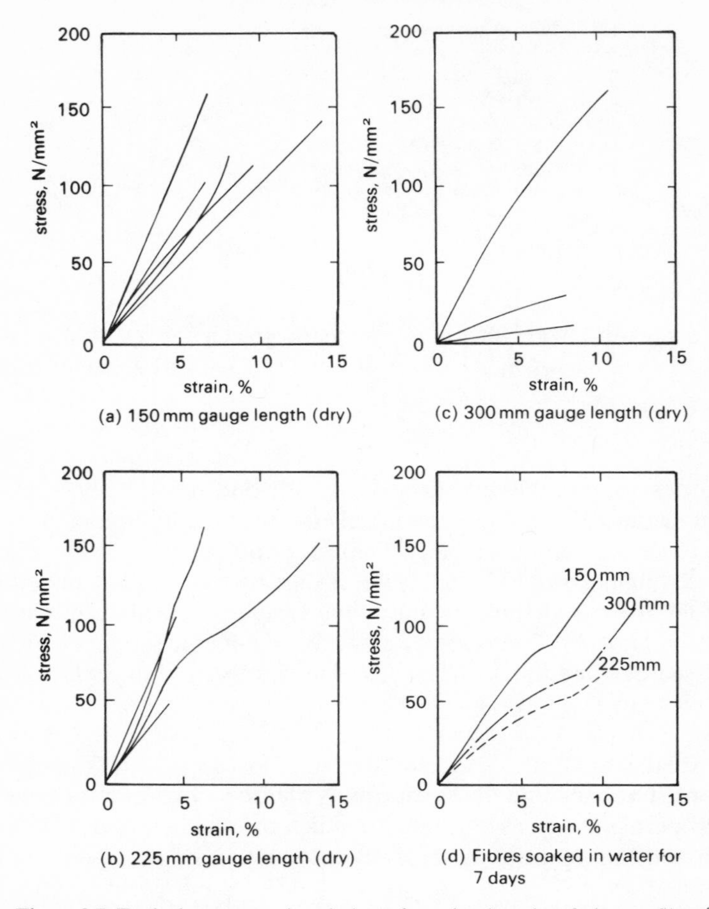

Figure 3.7 Typical stress–strain relation of varying lengths of akwara fibres[88]

three different lengths of akwara fibres[88]. The curves for the dry specimens are in general, predominantly linear and exhibit brittle failure characteristics. The soaked specimens exhibit two distinct stress–strain regions. After a certain strain, the fibres appear to have undergone some sort of strain-hardening. Values of the initial tangent modulus show wide scatter with considerable overlap in values between different lengths of fibres, values being in the range of 1 to 4 kN/mm^2.

Elephant grass is a tall, erect and stout perennial plant which is commonly found near watercourses, rivers and streams. It can grow as tall as 3 m, but 2.2 m is the common maximum size and the average diameter is 20 mm. The stem, being solid, contains a pith made of soft fibres. The crust is thin and fibrous and fibres are extracted mainly from the crust. The fibres are tough and sharp and therefore hand-extraction is difficult. The stress–strain character-

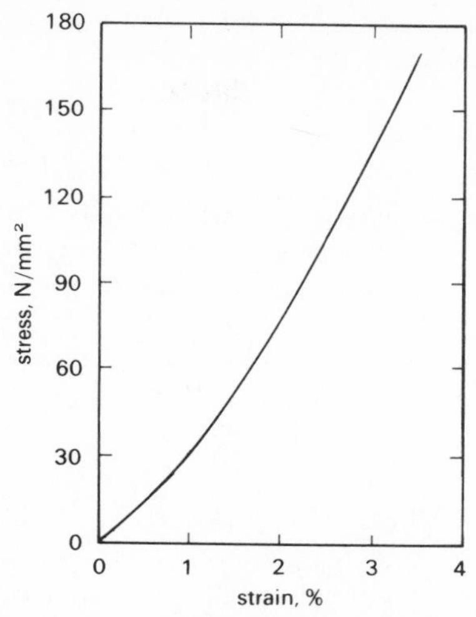

Figure 3.8 Typical stress–strain relation of elephant-grass fibres[91]

istic of elephant-grass fibre is shown in Fig. 3.8 and other properties in Table 3.7 which shows that the elephant grass fibre has the greatest tensile strength.

Water-reed plants are abundantly found on banks of rivers, streams, lakes and ponds in many countries. They generally grow in bushes, and when fully grown their height ranges from 2 to 3 m. The diameter of the matured stem may be as large as 20 mm. The stem consists of an empty interior and a strong fibrous crust about 5 mm thick. The fibres are extracted from the crust[89,100]. Fibre properties are given in Table 3.7 which shows the superiority of water-reed fibres with respect to modulus of elasticity.

Plantain is a tree-like tropical herbaceous plant allied to the banana family. The trunk is fibrous and fibres are easily extracted by hand. The fresh fibres are moderately strong and flexible[99] (Table 3.7).

Musamba is a hardwood tree abundantly grown in many countries. The bark of the tree is fibrous. Fibres are tough and moderately strong (Table 3.7)

Table 3.7 Physical and mechanical properties of some vegetable fibres[89]

Fibre type	*Tensile strength (N/mm^2)*	*Maximum strain (%)*	*Modulus of elasticity (N/mm^2)*	*Mean diameter (mm)*		*Minimum anchorage length (mm)*
				dry	*wet*	
Elephant grass	178	3.60	4936	0.45	0.41	25
Water reed	70	1.19	5193	1.10	0.98	20
Plantain	92	5.90	1436	0.43	0.39	30
Musamba	83	9.70	941	0.82	0.77	30

and extraction of fibres is very difficult. Musamba fibres are well-known for their use in making twines as a rope substitute[89,100].

3.2.10 *General remarks*

Of the natural fibres reported herein, coconut fibre, sisal fibre, bagasse fibre, bamboo fibre, jute fibre, flax fibre, akwara fibre, elephant grass fibre and wood fibre show greatest promise as reinforcing materials for concrete.

3.3 Production technology

3.3.1 *Objective*

The uniform dispersal of natural fibres in a cementitious matrix distributes stresses and improves microcracking. The principal objective in the production of natural-fibre reinforced concrete is therefore to distribute the fibres in such a way as to allow fibres to perform their desired functional role and to achieve a composite action between the fibres and the matrix. In addition, the technology should ensure that fibres are bonded well to the matrix. It is well established[1–9] that the most exploitable form of fibre-cement composites is to be made using short discontinuous fibres in either two-dimensional planar orientation, as in thin sections, or in random three-dimensional orientation in thick sections. Because of the particular nature of the cement constituent of the matrix, it is usually possible to incorporate only a small volume of discrete fibres in the matrix[101], and in any case, economic and other considerations dictate that the use of fibres be optimized.

3.3.2 *Matrix properties*

The constituent materials of natural-fibre reinforced concrete are small-diameter discontinuous discrete natural fibres of various origin and a matrix of cement, aggregates, water and admixture (if there is any). In fibre-reinforced concretes, the matrix binds the fibres together, protects them and takes part in the transfer of stresses to and from fibres. In general, the concrete matrices used with fibre reinforcement differ from conventional concretes in having a higher cement content, a lower coarse aggregate content and a small size of aggregates. The dimensional stability of the matrix can be improved by adding various inert fillers and pulverized fuel ash which have the effect of modifying the flow characteristics and other properties of the matrix[101–106].

3.3.3 *Procedure*

The four major steps in fibre-reinforced concrete production technology are fibre preparation, mixing of ingredients, placing and curing. Natural-fibre reinforced concrete construction, unlike other sophisticated engineering constructions, requires a minimum of skilled labour and utilizes readily available local materials. Proper attention should be paid to the control of the quality of construction otherwise the purpose of fibre-reinforced construction will not be served. The skills for natural-fibre reinforced concrete construction techniques are quickly acquired and the requisite quality control can be achieved using fairly unskilled labour for the fabrication under the supervision of skilled personnel. The most important advantage of these composites is that they can be fabricated into almost any desired shape to meet the needs of the user.

3.4 Properties

A true appreciation of the relevant properties of any material is necessary if a satisfactory end product is to be obtained and the natural-fibre reinforced concrete in this respect is in no way different from other materials. The performance of all types of natural-fibre reinforced concretes is very much influenced by the production process, quality control and the manner in which the material is placed in position.

Natural-fibre reinforced concretes constitute a new and distinct group of building materials which exhibit similar behaviour to those of conventional fibre-reinforced concretes produced from steel and other inorganic/synthetic fibres. As in conventional fibre reinforced concrete, the fibres act as crack-arresters which restrict the growth of flaws in the cement matrix from enlarging under stresses into visible cracks which ultimately cause failure. By restricting the growth of cracks, the usable tensile strength of the composite material is increased to a useful and predictable level. The dispersion of fibres in the brittle matrix offers a convenient and practical means of achieving improvements in many of the engineering properties of the material such as fracture, tensile and flexural strength, toughness, fatigue and impact resistance.

3.4.1 *Factors affecting properties*

Properties of natural-fibre reinforced concretes are affected by many factors. Some of the important factors are listed in Table 3.8. The list, whilst not exhaustive, emphasizes the complexity of production of a good-quality natural fibre reinforced composites. Fibre type, length and volume fraction have all been reported to have significant effects on its properties[49,60,70]. Optimum fibre length and volume fraction for most of the fibre types are around 25 mm and 3 percent respectively[106].

3.4.2 *Properties in the fresh state*

Fibre-reinforced mixes can be produced using power-driven pan and paddle mixers[101]. The constituents (cement and aggregates) are mixed first and then a

Table 3.8 Factors affecting properties of natural fibre reinforced concretes[106]

Factors	*Variables*
Fibre type	Coconut, sisal, sugarcane bagasse, bamboo, jute, wood, vegetables (akwara, elephant grass, water reed, plantain and musamba).
Fibre geometry	Length, diameter, cross-section, rings and hooked ends.
Fibre form	Mono-filament, strands, crimped, and single-knotted.
Fibre surface	Smoothness, presence of coatings.
Matrix properties	Cement type, aggregate type and grading, additive types.
Mix design	Water content, workability aids, defoaming agents, fibre content.
Mixing method	Type of mixer, sequence of adding constituents, method of adding fibres, duration and speed of mixing.
Placing method	Conventional vibration, vacuum dewatering sprayed-up concrete member, extrusion, and guniting.
Casting technique	Casting pressure.
Curing method	Conventional, special methods.

predetermined quantity of fibres of definite size is progressively added through a reciprocating-type fibre dispenser attached to the mixer. There are other methods of which the winding process[107] of producing fibre-reinforced concretes with exact properties has wider applications[9]. After mixing, operations such as transporting, placing, compacting and finishing of the fresh concrete can all considerably affect the properties of the hardened composite. It is therefore important that the fibres remain uniformly distributed and undamaged within the matrix during the various stages of the fabrications process, and that full compaction is achieved. When either of these conditions is not satisfied, the properties of the resulting composite are adversely affected. The workability of mixes containing a high volume fraction of natural fibres, however, requires careful design. Whilst the slump test is commonly used to assess the workability of conventional concretes, it is not generally suitable for natural-fibre reinforced concretes because many fibre reinforced mixes respond satisfactorily to vibration even though they have zero slump. With respect to aspect ratios for a given workability, a higher volume fraction of low aspect ratio fibres may be incorporated into the mix or vice-versa[101].

The incorporation of natural fibres into a mix decreases the workability and increases the void content due to entrainment of additional air[106–109]. The decrease in workability is basically due to the absorption, surface area, and especially, the size and shape of the fibres in relation to the other constituent particles in the mix. Unworkable mixes generally lead to non-uniform fibre distribution resulting in variation in properties between specimens from the same mix. The increase in void content is also due to the inadequate compaction of the unworkable mixes. The amount of fibres that can be added to a mix is limited by the phenomenon of 'balling'[27,35] where the fibres (if the fibres are long, i.e. have a high aspect ratio, greater than 100) have a strong tendency to intermesh and form fibre balls which cannot be easily separated. The balling of fibres results in an unworkable and segregated mix which ultimately produces a highly porous and honeycombed material. The balling of fibres, when large volume fractions are used, can be reduced by reducing the coarse aggregate content[20,35]. However, there is a limit to the volume of fibres that can be added to a mix beyond which the balling of fibres takes place and this mainly depends upon the nature and type of fibres, and the mix proportions.

Mixing methods have also been developed that minimize the balling problem to a great extent[30–35]. In order to improve workability, it is customary to increase the water/cement ratio of the mix at the expense of compressive strength or to use suitable admixtures which can improve the workability and strength properties of the concrete. The reduced workability can, of course, be exploited to produce concretes with different surface patterns to improve the appearance of the material. The low workability is also found useful during guniting where fibres help to hold the material together as the surface layers are built up. Moreover, in the fresh state, the fibres have the ability to hold together thin sheets of cement paste or mortar. These thin sheets can be suitably formed into folded or pressed shapes in much the same way that asbestos-cement sheets are moulded[106]. In most of the investigations of natural-fibre reinforced concretes, only short chopped fibres have been used. But it is also possible to incorporate continuous strands of natural fibres with or without rings or hooked ends in a cement paste or mortar to gain preferential directional properties from the fibres[91]. Such fibre strands can be used to make hollow posts and other suitable products. Long strands have also been used in the manufacture of products such as window panels and wall claddings from cement pastes and mortars. Another useful application of natural-fibre reinforced cement paste or mortar in the fresh state is the formation of shell and architectural structures by using a plastic membrane on which a thin layer of matrix can be placed or sprayed. All these indicate possible ways in which the properties of natural-fibre reinforced concretes in the fresh state can be exploited with economic advantages.

3.4.3 *Properties in the hardened state*

Important properties of the hardened fibre reinforced composite are strength, deformation under load, crack arrest, energy absorption, durability, permeability and shrinkage. In general, the strength is considered to be the most important property and the quality of natural-fibre reinforced concretes is judged mainly by their strength.

3.4.3.1 *Strength properties.* The ultimate strength depends almost entirely

Table 3.9 Effect of fibre volume fraction and mix proportions on modulus of rupture of sisal-fibre reinforced concrete[49]

Mix proportions cement: sand: coarse aggregate	*Water-cement ratio*	*Fibre length (mm)*	*Fibre volume fraction (%)*	*Curing period (days)*	*Specimen size (mm)*	*Modulus of rupture N/mm²*
1:1.8:2.4	0.57	50	3	28	100 × 100 × 500	3.70
1:1.8:2.4	0.57	50	3	35	100 × 100 × 500	3.80
1:0.78:0.82	0.36	50	5	14	100 × 100 × 500	4.80
1:3:0	0.50	50	2	7	80 × 80 × 300	3.70
1:3:0	0.50	50	1	28	80 × 80 × 300	3.30

Table 3.10 Effect of fibre volume fraction on strength, modulus of elasticity and Poisson's ratio of sugarcane bagasse-fibre reinforced cement composites[60]

Fibre length (mm)	*Fibre volume fraction (%)*	*Ultimate strength N/mm²*	*Modulus of elasticity N/mm² × 10³*	*Poisson's ratio*
Direct tension test				
	1	3.87	12.40	0.202
25	2	2.95	8.54	0.235
	3	2.43	6.40	0.248
Compression test				
	1	21.83	13.86	0.246
25	2	12.26	9.99	0.260
	3	6.26	6.78	0.281

upon the fibre type, length and volume fraction of fibres and also on the properties and proportion of other constituent materials (Tables 3.9–3.11 and Fig. 3.9). Figure 3.9 shows that tensile strength first increases with increasing fibre length, and after reaching a maximum value, the strength decreases for longer fibres mainly due to the bundling effect of the fibres. With the inclusion of short randomly-distributed natural fibres in concretes, various strength properties such as tensile, flexural and impact strength are substantially increased.

Mansur and Aziz[70] have investigated the stress–strain behaviour of jute-fibre reinforced cement mortar. Their results are shown in Fig. 3.10 which shows that the failure of specimens without jute fibres occurs suddenly at ultimate load and that the specimens break apart into two separate pieces in a brittle manner. The specimens with jute fibres, however, remain intact as one piece even after the maximum load is reached and continue to carry a significant amount of load in the post-maximum load stage. Similar behaviour

Table 3.11 Effect of fibre length and volume fraction on strength parameters of jute-fibre reinforced cement composites[70]

Mix ratio cement/sand	*Fibre volume fraction (%)*	*Fibre length (mm)*	*Compressive strength (N/mm^2)*	*Tensile strength (N/mm^2)*	*Modulus of rupture (N/mm^2)*	*Flexural toughness (Nmm)*	*Young's modulus (kN/mm^2)*	
							Compressive	*Tensile*
	0	—	31.44	1.20	2.81	35	15.54	9.50
	1	25	35.69	1.36	3.74	554	14.37	9.92
	2	25	30.00	1.96	4.50	875	12.30	11.60
1:0	3	25	37.44	2.08	3.82	740	12.84	11.20
	4	25	35.00	1.68	3.30	687	13.12	11.40
	2	12	30.57	1.72	3.88	651	15.20	10.40
	2	18	28.70	2.36	4.13	840	11.00	10.00
	2	38	31.18	1.75	4.40	827	11.81	9.20
	0	—	38.42	2.04	4.20	51	14.80	15.40
	2	12	37.43	2.51	5.62	878	14.30	18.00
1:1	2	18	32.44	2.18	5.02	1011	14.30	13.43
	2	25	32.76	2.16	4.50	1099	12.50	14.57
	2	38	27.97	2.12	4.00	817	8.50	17.70
	0	—	34.96	2.09	3.75	46	12.20	17.77
	2	12	27.97	2.33	4.44	824	10.00	22.86
1:2	2	18	28.72	2.48	4.60	1009	13.41	22.56
	2	25	32.46	2.03	3.92	797	14.22	16.00
	2	38	24.97	1.63	3.76	743	15.25	16.66

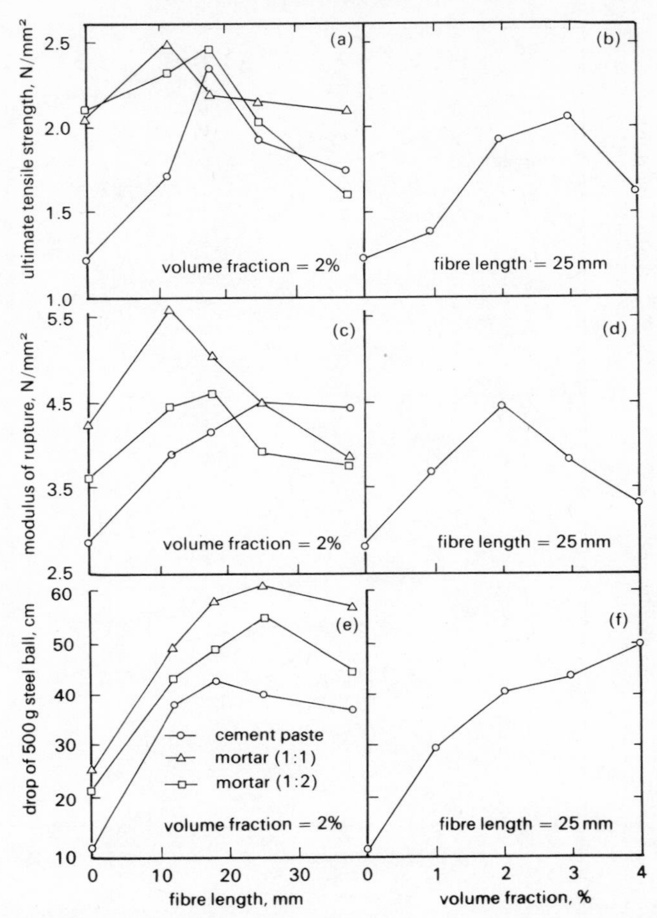

Figure 3.9 Effect of fibre length and volume fraction on tensile, flexural and impact strength of jute-fibre reinforced cement composites[70]

both in tension and flexure have been observed by Das Gupta *et al.*[37] and Paramasivam *et al.*[39] for coconut-fibre reinforced concrete materials, by Gram[47], Swift and Smith[49] for sisal-fibre reinforced concretes, by Krenchel and Jensen[110], Page *et al.*[111] Coutts *et al.*[75,83,84] and Andonian *et al.*[74] for wood-fibre reinforced concrete products. The addition of fibres thus converts the brittle matrix into a ductile material.

Extensive studies carried out by Pakotiprapha, Pama and Lee[63–65] on the behaviour of bamboo fibre-reinforced cement composite reveal that the post-cracking ductility of the composite is contributed by the bamboo fibres which serve as crack arresters. In addition, the post-cracking behaviour is considerably influenced by the volume fraction, fibre length and the mechanical

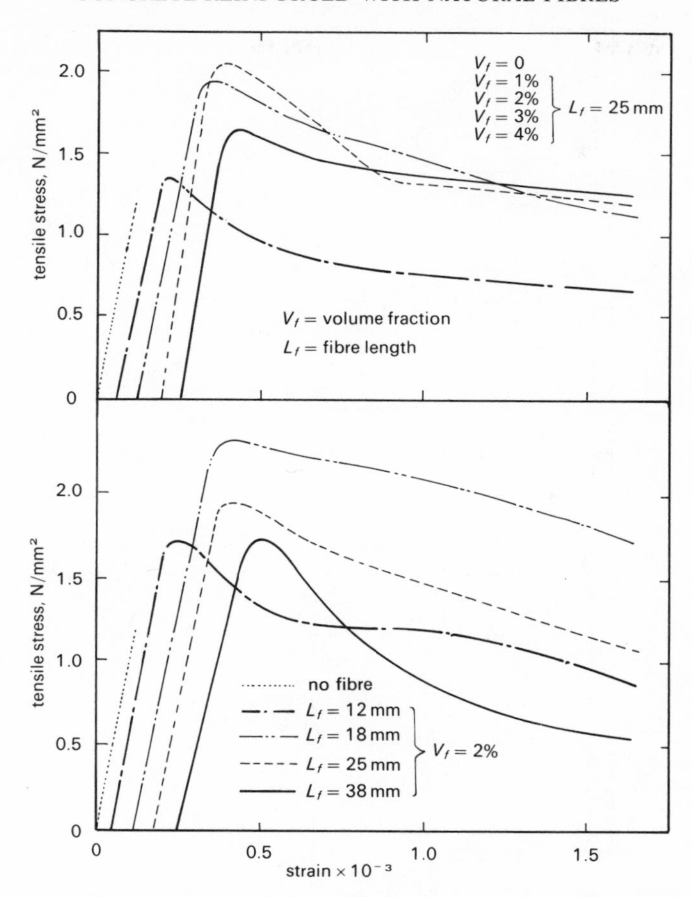

Figure 3.10 Stress–strain behaviour of jute-fibre reinforced cement composites[70]

properties of the fibre as well as the distribution of the interfacial bond stress between the fibre and the matrix.

The distinctive properties of natural-fibre reinforced concretes are thus improved tensile and bending strength, greater ductility, greater resistance to cracking and hence improved impact strength and toughness. The load-deflection characteristics of concretes using coconut fibres, sisal fibres, bagasse fibres, jute fibres, wood fibres and some vegetable fibres have been extensively studied by Das Gupta *et al.*[34,37], Paramasivam *et al.*[39], Gram[38,47,85], Racines and Pama[60], Mansur and Aziz[69–70], Andonian *et al.*[74], Coutts[76,86,87], Uzomaka[88], and Lewis and Mirihagalia[90,91] respectively. Some of the results are shown in Figs. 3.11–3.13. Paramasivam *et al.*[39] have observed (Fig. 3.11) that the slope of the load–deflection curves increases with increase in volume fraction and length of fibres. Similar behaviour was also observed by Racines and Pama[60], Mansur and Aziz[70] (Fig. 3.12) and Andonian *et al.*[74] (Fig. 3.13).

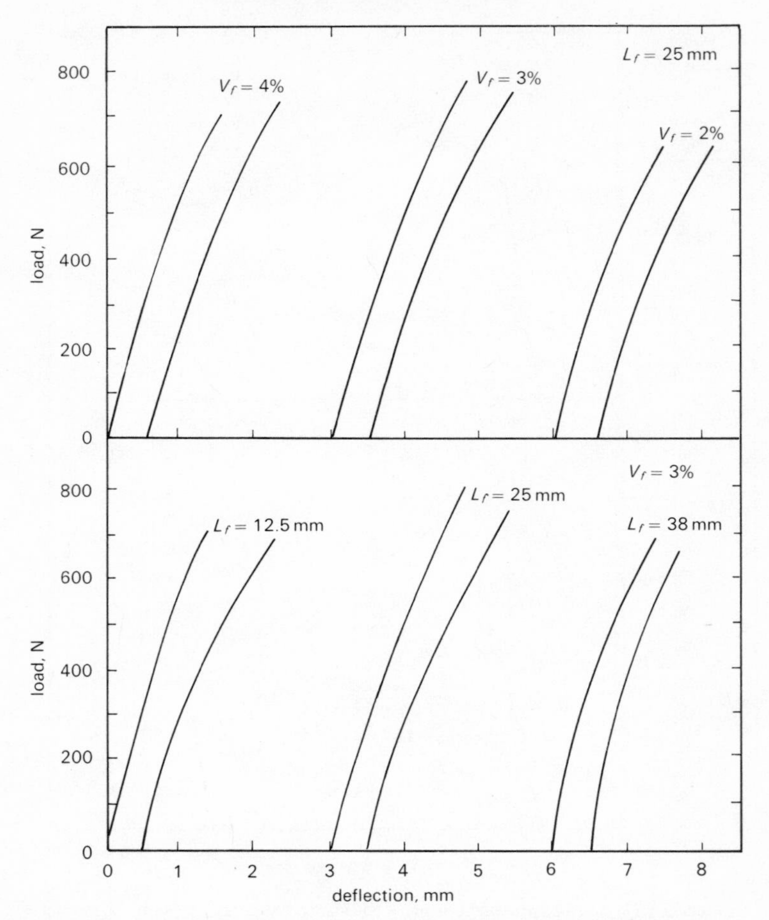

Figure 3.11 Load–deflection curves of coconut-fibre reinforced slabs[39]

In all cases, the observation of the failure surface of the specimens confirms that specimens generally fail by fibre pull-out rather than by fibre fracture. Poor bonding of the fibres is often due to swelling of the fibres in the wet mix and subsequent shrinkage upon drying. As suggested by Das Gupta *et al.*[34,37] and Cook *et al.*[36], the bond between the natural fibres and the matrix can be improved by applying a casting pressure resulting in an increase in strength (Fig. 3.14). The main effect of the casting pressure is to reduce the void space and to densify the concrete. Paramasivam *et al.*[39] observed that a casting pressure of 1.5 atmospheres increased the flexural strength of coconut-fibre reinforced corrugated slabs by 20% in comparison to those produced without applying casting pressure.

In addition to improvements in strength, a considerable increase in toughness is also imparted by the inclusion of natural fibres in the composite,

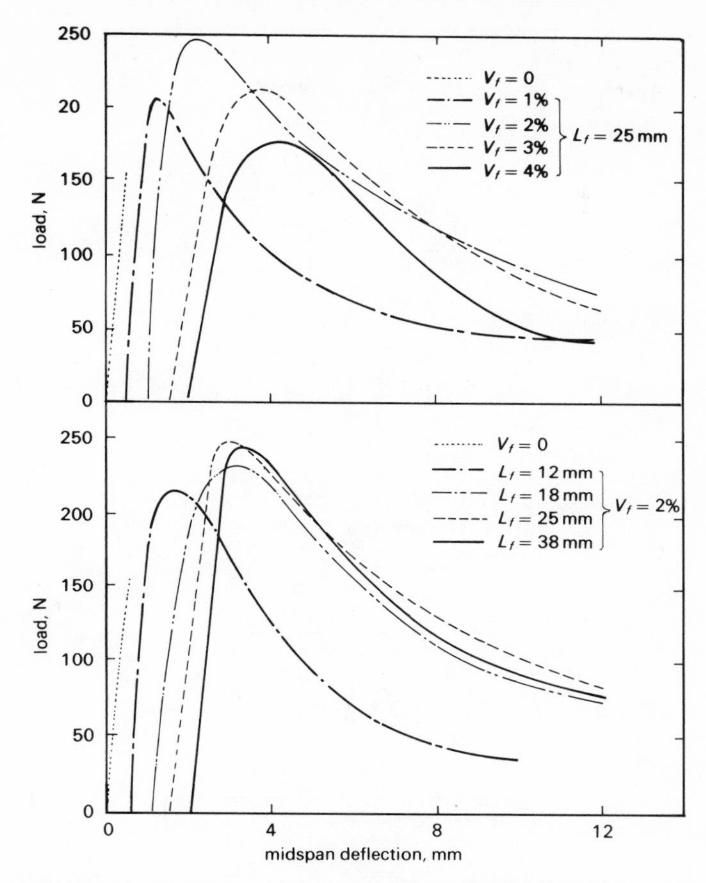

Figure 3.12 Load–deflection curves of jute-fibre reinforced cement composites[70]

as shown in Table 3.11. The increase has been found to be substantial which indicates the ability of a structural element to absorb large amounts of energy prior to failure[70]. Another related property of these natural fibre composites is their resistance to impact. Mansur and Aziz[70] have reported that the inclusion of jute fibres (length, 25 mm and volume fraction, 3%) in a cement paste improves the impact strength considerably, the maximum increase being 400%. Similar improvement of impact strength has been observed by Siraskar and Kumar[112] in their study of shatterproofness of both coconut and jute fibre reinforced cement composites (Table 3.12). Swift and Smith[113] have also reported the superiority of sisal-fibre reinforced concrete as an earthquake-resistant construction material.

3.4.3.2 *Crack arrest and energy absorption properties.* Natural-fibre reinforced composite behaves as a homogeneous material within certain limits. The

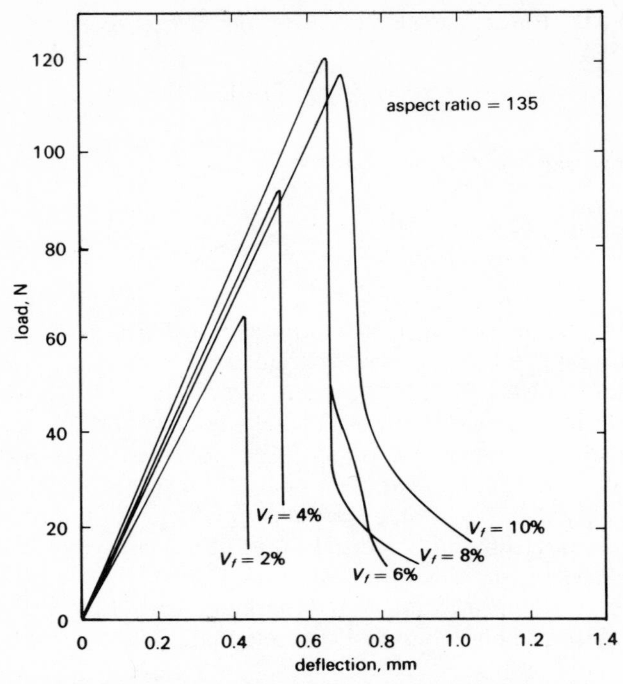

Figure 3.13 Load–deflection curves of wood-fibre reinforced cement composites[74]

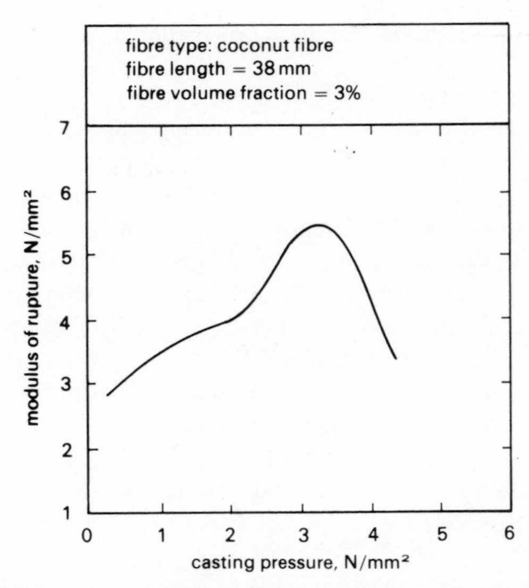

Figure 3.14 Effect of casting pressure on modulus of rupture of coconut-fibre reinforced cement composites[34]

Table 3.12 Impact strength of various natural-fibre reinforced concretes[112]

Concrete type	*Impact strength, N.m/m²*		
	curing periods, days		
	7	*21*	*30*
1. Plain concrete	1.35	2.03	6.08
2. Concrete with 3% coconut fibre	4.73	11.48	20.93
3. Concrete with 3% jute fibre	2.70	4.73	20.25

random distribution and high surface-to-volume ratio (specific surface) of the fibres results in a better crack-arresting mechanism[34,65,70,106]. With low fibre contents that are normally used in cement composites (from 2 to 4% by volume), the strain at which the matrix cracks is little different from that observed in plain cement pastes, mortars or concretes. However, once cracking has started, the fibres act as crack-arresters, and also absorb a significant amount of energy if they are pulled out from the matrix without breaking. The inclusion of short natural fibres in cement-based composite materials, nevertheless, increases the first crack strength[20,34,65]. These two properties, crack-arresting mechanism and energy absorption as the fibres pull out, make the natural-fibre reinforced concrete materials tough and enable them to withstand impact far more satisfactorily than plain concretes (Tables 3.11 and 3.12). These two properties have made natural fibre composites useful in small precast products where accidental damage from impacts creates large waste. Further, once the matrix has cracked, the fibres carry a major portion of the tensile stress in the composite material.

3.4.3.3 *Thermal properties.* Thermal properties of natural-fibre cement composites is important in its performance over long periods under varying environmental conditions. Basic parameters involved are thermal conductivity, thermal diffusivity, specific heat and coefficient of thermal expansion. Of these, the first three are interrelated. Table 3.13 gives the thermal conductivity values reported by Paramasivam *et al.*[39] for coconut-fibre reinforced slabs.

Table 3.13 Thermal conductivity of coconut-fibre reinforced corrugated slabs[39]

Cement–sand ratio	*Water–cement ratio*	*Fibre length (mm)*	*Fibre volume volume (%)*	*Slab thickness (mm)*	*Thermal conductivity (W/m°K)*
			2		0.62
			3		0.63
1:0.5	0.35	25		25.4	
			4		0.61
			5		0.68

Table 3.14 Comparison of properties of elephant-grass fibre reinforced roofing sheets with those reinforced with asbestos fibres[90]

Properties	*Cement sheets reinforced with elephant-grass fibres*	*Cement sheets reinforced with asbestos fibres*
1. Consistency at 25% water content	15%	11%
2. Impact strength	2.82 J (SFS* = 3.1)	4.04 J (SFS = 20.3)
3. Flexural strength	10.5 N/mm^2 (SFS = 5.10)	17.8 N/mm^2 (SFS = 20.0)
4. Impermeability	excellent	excellent
5. Water absorption	16.3%	20.6%
6. Coefficient of thermal conductivity	0.33 W/m°K	0.36 W/m°K
7. Sound transmission of 833 Hz signal	22% when dry 30% when wet	26% when dry 40% when wet
8. Combustibility (BS 476-Part 4)	Non-combustible	Non-combustible
9. Linear expansion	0.22%	0.24%
10. Density	1,770 kg/m^3	1,540 kg/m^3

*SFS = specific surface area = surface area of fibre per unit surface area of sheet 6 mm thickness containing 1 kg cement

These values compare very well with those of asbestos-cement boards as reported by Eldridge[114]. Lewis and Mirihagalia[90] also made a comparison of thermal conductivity and linear expansion of elephant-grass fibre reinforced cement roofing sheets with those of asbestos-fibre reinforced sheets (Table 3.14). It is seen that these properties of natural fibre reinforced concretes are comparable to those of asbestos cement sheets.

3.4.3.4 *Acoustic properties.* Acoustic properties of construction materials are of great importance in their use in buildings. Basically, two acoustic properties—sound absorption and sound transmission—are important. The sound transmission values as observed by Lewis and Mirihagalia[90] in their study of the properties of elephant-grass fibre reinforced roofing sheets are given in Table 3.14. The sound absorption coefficient values of coconut-fibre reinforced slabs studied by Paramasivam *et al.*[39] are given in Table 3.15 which shows a wide range of values for the sound absorption coefficient i.e. 2.5 to 8% for low-frequency range and 12 to 18% for high-frequency range. The volume fraction of fibre and the thickness of the material do not have significant effect on the sound absorption coefficient but the surface finish plays an important role. The variation of sound absorption with frequency is shown in Fig. 3.15 for a typical coconut-fibre reinforced corrugated board specimen 10 mm thick with 3% fibre volume fraction[39]. The absorption coefficient does not vary significantly at the lower frequency range (less than 100 Hz) but it increases considerably with increasing frequency. Asbestos cement boards are also reported to have a sound absorption coefficient in the

Table 3.15 Sound absorption coefficient* of coconut fibre reinforced slabs[39]

Specimen	*Frequency (Hz)*	*Sound absorption coefficient (%) for* 125	250	500	1000	2000	4000
Fibre volume fraction (%)	*Thickness (mm)*						
2	13	2.5	4.5	6.0	3.5	13.5	48.0
	10	2.5	6.0	6.0	5.5	12.0	44.0
	5	3.0	5.0	2.5	3.5	10.5	15.0
3	13	3.5	4.5	5.5	5.5	18.5	27.5
	10	3.0	4.5	5.5	6.0	16.0	45.0
	5	3.0	4.0	2.5	5.5	14.0	26.0
4	13	4.0	4.5	3.0	6.0	22.0	23.0
	10	4.0	5.0	4.0	13.5	19.5	31.0
	5	4.5	5.0	7.0	7.5	14.0	43.0

*Values are as percentages

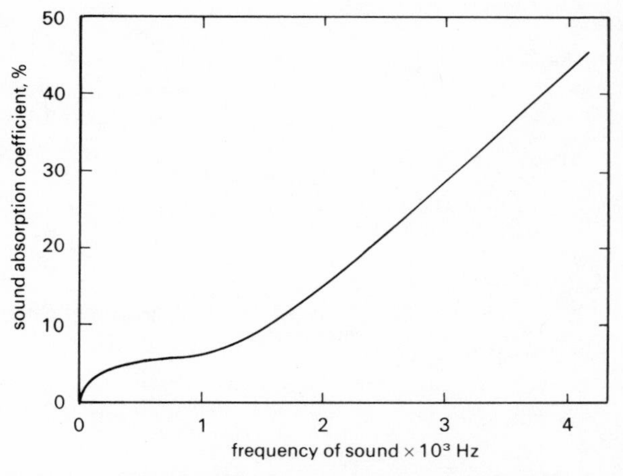

Figure 3.15 Sound absorption coefficient with respect to frequency of coconut-fibre reinforced slabs[39]

range of 5 to 8% at higher frequency depending upon thickness[115]. Natural-fibre reinforced composites, therefore, appear to provide inherently better sound insulation with respect to sound transmission[39,57,88,92].

3.4.3.5 *Permeability, water absorption, swelling and shrinkage characteristics.* Permeability of fibre-reinforced concrete materials is of interest in relation to watertightness of liquid-retaining and other structures. Moreover, the ingress of moisture into concrete due to permeability affects its thermal insulation properties. The permeability of natural-fibre reinforced concretes has been investigated by many researchers[38,39,40,47,82] and typical values are about 3×10^{-6} cm/sec depending upon the nature of the fibre, fibre volume fraction and casting pressure. Water absorption is normally in the range of 15% to 30% but with the application of surface paints, the value can be reduced to 2 to 7% (Tables 3.16–3.18). Some natural fibres absorb a significant amount of water from the wet matrix resulting in swelling and shrinkage as the

Table 3.16 Summary of test results of water absorption, linear expansion, shrinkage, impact resistance and permeability of bagasse fibre reinforced cement composites[59,60]

Water absorption, %

Fibre volume fraction	*Plain bagasse fibre-cement*	*With admixture*	*With cement paste surface coating*	*With paint surface coating*
0.1	19.11	16.47	16.20	2.04
0.2	28.05	26.10	25.18	5.23
0.3	33.76	31.53	29.32	6.69

Expansion, shrinkage, impact resistance and permeability

Fibre volume fraction	*Maximum linear expansion (%)*	*Equilibrium shrinkage (%)*	*Impact resistance (index number)*	*Permeability (cm/sec $\times 10^{-6}$)*
0.1	0.050	0.045	23	3.868
0.2	0.124	0.082	18	7.742
0.3	0.144	0.113	16	11.140

Table 3.17 Comparison of physical properties of coconut fibre reinforced roofing sheets and asbestos roofing sheets[31]

Characteristics and properties	*Coconut-fibre reinforced roofing sheets*	*Asbestos-cement roofing sheets*
1. Pitch of corrugation, mm	146	146
2. Depth of corrugation, mm	48	48
3. Length of sheets, m	1.5–2.0	1.5–3.0
4. Width of sheets, m	1.0	1.05
5. Weight, kg/m^2	11.5–12.0	13.5
6. Breaking load for a span of 60 cm, N/m	50	—
7. Breaking load at a span of 100 cm, N/m	19	50
8. Thermal conductivity, kcal/cm/m^2	0.09	0.24
9. Water permeability through finished surface in 24 hours	almost nil	—
10. Acid resistance as per I.S.: 5913–1970, N/mm^2	9.30×10^3	9.26×10^3

Table 3.18 Comparison between properties of surgarcane bagasse fibre reinforced cement boards and specified values for building boards in ASTM C208-72[59,60]

		Test values		
	Specified value on	*Fibre cement ratio (by wt.)*		
Property	*ASTM C208-72*	*0.2*	*0.3*	*0.4*
1. Transverse load, N	Min. 54.43	185.00	193.80	149.80
2. Deflection at specified minimum load, mm	Max. 21.59	0.89	1.09	1.85
3. Modulus of rupture N/mm^2	Min. 16.87	2.84	2.87	2.27
4. Tensile strength parallel to surface N/mm^2	Min. 10.55	8.16	8.69	—
5. Linear expansion, %	Max. 0.50	0.07	0.09	0.10
6. Water absorption, %	Max. 7.00	5.23	6.90	—

matrix hardens. The use of cheap water-sealing agents has, however, been found to be effective in reducing the water absorption potential of natural fibres thereby improving the bond strength[47,70,74,91].

3.5 Durability

Besides its ability to sustain loads, natural-fibre reinforced concrete is also required to be durable. Durability relates to its resistance to deterioration resulting from external causes (effects of environmental and service conditions such as weathering, chemical actions and wear) as well as internal causes (effects of interaction between constituent materials such as alkali-fibre reaction, volume changes, permeability and absorption). In order to produce a durable fibre-reinforced concrete, care should be taken to select suitable constituent materials in adequate quantities and in suitable proportions and good quality natural fibres of specified length and volume fraction for producing a homogeneous and fully compacted fibre-reinforced concrete mass.

The successful performance of natural-fibre reinforced concrete composites is dependent more on their durability against the environment in which they are exposed than on their strength properties. It is therefore essential that such a concrete should withstand the conditions for which it has been designed without deterioration over a period of years. Limited data exist at present on the durability aspects of natural fibre reinforced concrete in different environmental conditions. Poor dimensional stability of natural fibres due to moisture changes has given rise to durability problems and various proposals to provide protective treatment of the fibres are presented in literature[47,76,106]. Several investigations have noted the embrittlement of sisal, coir and jute fibre reinforced concrete. The reason for such embrittlement has been found to be the alkaline pore-water in the concrete which dissolves the fibre components so that they become decomposed and lose their reinforcing capacity. Extensive studies by Gram[117] at the Swedish Cement and Concrete Research Institute have shown that the best way of counteracting the embrittlement of these fibres is to reduce the alkalinity of the pore-water. This can be achieved by replacing 40 to 50% of the cement content by silica fume. The use of high alumina cement also reduces the alkalinity of the pore-water and thus slows down the rate of embrittlement. Sealing the concrete pores with wax and resin and using various fibre-impregnating agents have also reduced the embrittlement to a satisfactory extent[47,76].

Gram[47] has reported extensive test data on the post-cracking strength of sisal-fibre reinforced cement mortar against alkali attack. The binder of the test specimens consisted of high-alkaline portland cement with about 1.6% free alkali. The fibres were uniformly distributed and aligned parallel to the mould. They were used either unprotected or protected by impregnation in formine and stearic acid. In addition, as a protective measure against alkali attack, a modified binder was used by replacing part of the portland cement with silica fume by 20% to 45% by weight. Two accelerated ageing tests were used. The composite specimens were stored in hot water at 50°C for different periods or subjected to alternate wetting and drying. The results are shown in Table 3.19.

When subjected to high-temperature curing, the test specimens showed high flexural residual strength varying from 11 to 21 N/mm^2 at ages of 1 to 2 years. Translated in terms of natural weathering, this corresponds to a safe service life of about 27 years in the tropics or over 400 years in the Swedish climate. This of course does not correspond to reality as there is evidence that specimens become very brittle in tropical climates within one year.

When subjected to alternate wetting and drying cycles, sisal fibres impregnated with formine and stearic acid gave a more durable composite

Table 3.19 Residual flexural strength in N/mm^2 of sisal-fibre concrete after accelerated ageing[47]

Ageing in hot water at 50°C

Test no	*Binder*	*Storage in days at 50°C*				
		28	91	180	365	730
72–98	OPC	13.8	11.2	13.5	11.5	12.1
135–158	30% SiO_2	17.7	16.5	16.3	17.3	20.6

Ageing by alternate wetting and drying

Test no	*Binder*	*Fibre*	*No. of cycles*			
			0	12	60	120
99–110	OPC	Unimpregnated	32.8	2.8	0.9	0.4
111–122	OPC	Impregnated	13.6	9.7	9.5	5.6
123–134	20% SiO_2	Unimpregnated	28.7	21.4	7.8	5.3
159–170	35% SiO_2	Unimpregnated	36.7	33.8	21.7	19.6
171–182	45% SiO_2	Unimpregnated	33.8	31.0	29.9	31.1

than the unimpregnated fibres*. The results show further that sisal-fibre reinforced composites become more durable when part of the portland cement is replaced by highly-active ultra-fine silica fume (SiO_2). The presence of silica fume reduces the alkalinity of the matrix, and at 45% replacement, dramatic improvements in residual flexural strength were observed, and over 90% of the original strength was retained after 120 cycles of wetting and drying (Table 3.19). These results show that the embrittlement of natural fibres can be delayed substantially by protective impregnation, and that further such embrittlement can be avoided almost completely by reducing the alkalinity of the cement matrix. The test method is also important, and accelerated wetting and drying appears to be more critical in establishing the resistance to alkali attack than curing at high temperature.

Various material performance properties like permeability, water absorption, thermal expansion, and shrinkage usually vary with the type of fibre used and increase with increasing fibre concentration. These can be significantly reduced if the surface is coated with suitable paints or admixtures[60,117]. The durability of natural-fibre reinforced concrete can therefore be enhanced by using rich mixes with suitable admixtures and coating the surface of the structural members by some approved paints. There is, however, a large volume of information available to date for improving the durability or providing protective measures for particular applications of natural-fibre reinforced concrete.

3.6 Practical applications

Potential applications of natural-fibre reinforced concretes for large-scale structural purposes are limited. Only in a few special applications are they practically and economically justified. One of the most promising fields for their application is that of composite construction in which the natural-fibre reinforced cement, mortar or concrete forms an outer layer to some other materials[118–120]. The outer layer can serve as a permanent strong and tough covering over a weak core and can provide architectural and ornamental features. The outer layer can also increase the strength or effective strain capacity of the tensile surface of a concrete beam and so reduce crack widths and spacings, and thereby provide a measure of surface crack control[119].

Natural-fibre reinforced concrete products like sheets (both plain and corrugated) and boards are light in weight and are ideal for use in roofing, ceiling, and walling for the construction of low-cost houses. Sisal-fibre reinforced concrete tiles, corrugated roofing sheets, pipes, gas tanks, water tanks, and silos are being extensively used in some African countries[54]. Products also include certain bituminous roofing sheets and damp-proofing felts (BS 747: 1977 and BS 743: 1970). Other natural-fibre reinforced materials include pitch-fibre pipes and conduits. Wood- and sisal-fibre reinforced

*The post-cracking flexural strength after 120 cycles was over 5 N/mm^2 compared to about 0.4 N/mm^2 for the unimpregnated fibres.

concrete panels and building boards are being used in Australia[47,82,83]. In Zambia, elephant-grass fibre reinforced mortar and cement sheets are being used for low-cost housing[121]. Natural-fibre reinforced cement composites have also found applications in ceiling and partition linings, eaves, soffits, sound and fire insulation[71–87]. Their special usages include applications where energy is to be absorbed or where impact damage is likely such as shatter- and earthquake-resistant construction, factory floors and foundations for machinery where accidental impact, shock and vibratory loads occur. Other conventional applications include rafts and beams for cellular foundation, pavements, slabs, aprons, and various types of shell structures[122]. All potential applications of natural-fibre reinforced concretes depend, of course, on the ingenuity of the designer and the builder taking advantage of the static and dynamic strength parameters, energy-absorbing characteristics, material performance properties, acoustic and thermal behaviour.

3.7 Concluding remarks

Although there has been considerable interest in the development of natural-fibre reinforced concrete materials in recent years, yet it is not easy to show fields in which these new materials have become an automatic choice. The reasons for the slow acceptance of these new types of fibre-reinforced materials are not difficult to understand. Conventional concrete is a cheap material that can be made with unsophisticated equipment from indigenous materials. Experience with steel, glass and other inorganic fibres gained over the past several years has assured the confidence of building fibre-reinforced concrete elements that are strong, durable, economic and of attractive appearance. Natural-fibre reinforced materials are relatively new, having unproven long-term durability. These materials, however, possess a number of useful attributes and it is in such areas where these attributes can be utilized that natural fibre composites can be exploited for wider use.

The greatest advantage of natural-fibre cement composites is, however, that the fibres are available in a variety of forms in abundance, freely, in the very countries where the need for housing and other low-cost construction is the greatest. It is tempting, but unrealistic, when considering natural-fibre cement composites, to compare their properties with other high-technology composites based on steel, glass and other organic fibres. With over half the population of the Third World living in slums, shanty towns and uncontrolled settlements, even cement, a relatively cheap building material in the developed world, is too expensive a material to be used for construction. The emergence of natural fibres and a wide range of modified cement matrices incorporating waste materials such as fly ash and rice husk ash provides an exciting challenge to the construction industry for housing, for temporary refugee camps, for providing roofing sheets for protection from the elements, in sanitation, water-supply and a host of other basic needs of humanity. Natural-fibre cement composites thus pose the challenge and the escape for a combination of unconventional building materials and traditional construction methods.

Natural fibre reinforced concrete is a new construction material. Until recently, most of its practical applications appear to have been based on the concept of improved strength properties. In reality, the role of this new construction material would not only be in the improvement of static strength but also to enhance resistance to cracking, improve ductility for better energy absorption and resistance to impact and shock loadings. If engineers and builders acknowledge these beneficial aspects of this distinct group of new materials, their potential uses are not only assured but also bound to be wide and varied.

From a critical appraisal of the forms and properties of natural-fibre reinforced composites, it is clear that the appropriate technology for a quality product development is of prime importance. For this, extensive investigation work needs to be devoted to the following fields: (i) classification of natural fibres with regard to physical and chemical properties, (ii) fibre morphological properties-interaction between fibres and the cement matrix, (iii) composite studies including both long-term and accelerated durability tests and (iv) development of product types and production methods including spraying techniques. To be more specific, further development work for economic and efficient methods of fibre extraction, pretreatment of fibres, conversion into usable forms, dispersion in to the cement matrix, fabrication and curing are desirable. Standard test procedures must also be established to evaluate product quality control. Design procedures must be standardized. Parameters such as strength, deflection, impact and abrasive resistance, water absorption, shrinkage, chemical resistance, acoustic requirements, thermal performance and durability must be properly assessed. Favourable areas of exploitation which depend upon the identification of problems or components where the special characteristics of this type of concrete materials can be used, must be singled out. A full functional specification must be prepared for each type of fibre and also for each particular application.

Acknowledgements

The authors extend their gratitude to all the publishers and authors for diagrams and tables referred to in this chapter.

References

1. The Concrete Society (1973) *Fibre-reinforced cement composites.* Technical Report 51.667, 1–77.
2. Swamy, R.N. and Fattuhi, N.I. (1974) 'Mechanics and properties of steel fibre reinforced concrete', *Proc. 1st Australian Conf. Materials*, Univ. of New South Wales, Sydney, 351–68.
3. ACI Committee 544 (1974) *Fibre Reinforced Concrete.* Publication No. SP-44, American Concrete Institute, Detroit, U.S.A.

4. Swamy, R.N. (1975) Fibre reinforcement of cement and concrete *RILEM Mater. Struct.* **8** (45), 235–54.
5. RILEM Symposium 1975, *Fibre Reinforced Cement and Concrete.* Ed. A.M. Neville *et al.*, Construction Press Ltd, Lancaster, England, **1**, 1–459.
6. Majumdar, A.J. and Swamy, R.N. (1977) 'Fibre Concrete materials'. Report prepared by RILEM Technical Committee 19-FRC, *RILEM Mater. Struct.* **10** (56). 103–20.
7. ACI Committee 544 (1978) Measurement of properties of fibre reinforced concrete. *ACI J.* **75** (7), 283–9.
8. Majumdar, A.J. and Laws, V. (1979) Fibre cement composites: research at BRE. *Composites* **10** (1), 17–27.
9. Shah, S.P. (1981) Fiber-reinforced concretes: a review of capabilities. *Concr. Constr.* 261–68.
10. George, J. and Joshi, H.C. (1959) Complete utilization of coconut husk, Part 1—Building boards from coconut husk. *Indian Coconut J.* **12** (2), 46–51.
11. Iyengar, N.V.R. *et al.* (1961) Thermal insulation board from coconut husk and pith. *J. Sci. Industr. Res.* **200**, 276–9.
12. George, J. and Bist, J.S. (1962) Complete utilization of coconut husk, Part IV—Three layer particle board with coconut husk particle core. *Indian Pulp and Paper* **16** (7), 437.
13. George, J. and Shirsalkar, N.M. (1963) Particle boards from coconut husk. *Research and Industry, (India)* **8** (2), 129–31.
14. Shirsalkar, M.M. *et al.* (1964) Fire resistant building boards from coconut coir. *Research and Industry, (India)* **9** (12), 359–60.
15. George, J. (1970) Building materials from coconut husk and its by-products. *Coir, (India)* **14** (2), 19–44.
16. Jain, R.K., and George, J. (1970) Utilization of coconut pith for thermal insulating concrete *Coir, (India)* **14** (3), 25–31.
17. Sridhara, S., Kumar, S., and Sinare, M.A. (1971) Fibre-reinforced concrete, *Indian Concr. J.* **45** (10) 428–30.
18. Siraskar, K.A., and Kumar, S. (1972) 'Fibre reinforcement for shatter proofness of concrete', *Proc. Symp. Modern Trends in Civil Eng.*, Univ. Roorkee, India, 214–219.
19. Marotta, P. (1973) *Tests on panels of coconut fibres impregnated with Portland cement* Division of Building Research-CSIRO, Australia, Report 2.
20. Singh, S.M. (1974) Agro-industrial wastes and their utilization. *Research and Industry (India)* **19** (2), 159–62.
21. Singh, S.M. (1975) Corrugated roofing sheet and building panels from coir wastes. *Coir (India)* 3–6.
22. Singh, T.P. (1975) *Structural properties of fibre reinforced concrete.* M.Sc. Thesis, Dept. of Civil Eng., Univ. Roorkee, India.
23. Saylor, I., Ball, G.L., Usmani, A.M., Werkmeister, D.W., and Falconer, J.R.P. (1975) *Development of low cost roofing from indigenous materials in developing nations.* 2nd Annual Report, Agency of International Development, US Department of State, Contract No. AID/CM/ta-C-73-12, Dayton.
24. St. John, D.A., and Kelley, J.M. (1975) The flexural behaviour of fibrous plaster sheets. *Cem. Concr. Res.* **5**, 347–62.
25. Zonsveld, J.J. (1975) 'Properties and testing of concrete containing fibres other than steel', *RILEM Symposium* 1975*: Fibre Reinforced Cement and Concrete.* The Construction Press Ltd, Lancaster, England, **1**, 217–26.
26. Mohan, D., and Singh, S.M. (1976) 'Low cost constructional material from coconut husk and rice husk', *Proc. Int. Symp. New Horizons in Construction Materials*, Envo. Publishing Co. Inc. **1** (1–3), 477–87.
27. Slate, F. (1976) Coconut fibres in concrete. *Eng. J. Singapore* **3** (1), 51–4.
28. Pama, R.P., Cook, D.J., and Oranratnachai, A. (1976) 'Mechanical properties of coir-fibre boards', *Proc. Conf. New Horizons in Construction Materials*, Envo Publishing Co. Inc., 391–404.
29. Singh, S.M. (1977) *Roofing sheets and building panels from coconut husk and fibres.* Hard Fibre Research Series No. 20, UN-FAO, New York.
30. Mohan, D., and Rai, M. (1977) 'New building materials and techniques for rural housing', *Proc. Int. Conf. on Low Income Housing Technology and Policy*, Asian Institute of Technology, Bangkok, 1077–93.

31. Weerasinghe, H.L.S.D. (1977) *Fundamental study on the use of coir-fibre boards as a roofing material.* M. Eng. Thesis, Asian Institute of Technology, Bangkok.
32. Jain, R.K., and George, J. (1978) *Coconut fibre cement concrete for thermal insulation.* Central Building Research Institute, Roorkee, India, 9–13.
33. Ganeshalingam, R., Paramasivam, P., and Nathan, G.K. (1981) An evaluation of theories and a design method of fibre-cement composites. *Int. Cem. Comp. Lightweight Concr.* **3** (2), 103–13.
34. Das Gupta, N.C., Paramasivam, P., and Lee, S.L. (1978) Mechanical properties of coir reinforced cement paste composites, *Housing Science* **2** (5), 391–406.
35. Singh, S.M. (1978) 'Coconut husk—a versatile building material', *Proc. Int. Conf. Mat. Constr. Dev. Count.*, Bangkok, 1, 207–20.
36. Cook, D.J., Pama, R.P., and Werasinghe, H.L.S.D. (1978) Coir fibre reinforced cement as a low cost roofing material, *Building and Environment* **13** (3), 193–8.
37. Das Gupta, N.C., Paravasivam, P., and Lee, S.L. (1979) 'Coir reinforced cement paste composites'. *Proc. Conf. Our World in Concrete and Structures*, Singapore, 111–6.
38. Gram, H-E. (1980) *Coir reinforced concrete.* Swedish Cement and Concrete Research Institute, Stockholm, Consultant Section, Report 7925, 1–9.
39. Paramasivam, P., Nathan, G.K., and Das Gupta, N.C. (1984) Coconut fibre reinforced corrugated slabs. *Int. J. Cem. Comp. Lightweight Conc.* **6** (1), 19–27.
40. Nilson, L. (1975) *Reinforcement of concrete with sisal and other vegetable fibres.* Swedish Council for Building Research, Document DIU, Svensk Byggtjanst, Stockholm, 1–68.
41. Treiber, E. (1977) *Properties of sisal fibres–a study of the literature.* Project 5527, Swedish Forests Products Research Laboratory, Stockholm, 1–23.
42. BRU Data Sheet. (1978) *Roof sheets made of sisal reinforced concrete.* Building Research Unit, P.O.Box 9344, Dares-Salaam, Tanzania, 1–8.
43. Cappelen, P. (1978) *Roof sheets made of sisal reinforced cocnrete.* Building Research Unit, Ministry of Lands, Housing and Urban Development, Tanzania, Working Report WR14, 1–7.
44. Gram, H-E., and Skarendahl, A. (1978) *A sisal reinforced concrete: Study No. 1 on material.* Swedish Cement and Concrete Research Institute, Stockholm, Consultant Section, Report 7822, 1–15.
45. Persson, H. (1978) *Sisal reinforced concrete: Study No. 1 on prototypes.* Div. Building construction, Royal Institute of Technology, Sweden, Report 16/1/1978, 1–3.
46. Persson, H., and Skarendahl, A. (1978) *Sisal reinforced concrete: Study No. 2 on prototypes.* Swedish Cement and Concrete Research Institute, Stockholm, Consultant Section, Report 7829, 1–9.
47. Gram, H.E. (1983) *Durability of natural fibres in concrete.* Research Report 1, Swedish Cement and Concrete Research Institute, Stockholm, p. 255.
48. Skarendahl, A. (1978) *Sisal reinforced concrete: Study No. 2 on materials.* Swedish Cement and Concrete Research Institute, Stockholm, Consultant Section, Report 7827, 1–2.
49. Swift, D.G., and Smith, R.B.L. (1978) 'Sisal fibre reinforcement of cement paste and concrete', *Proc. Int. Conf. Mat. Constr. Dev. Count.*, Bangkok, 1, 221–34.
50. Mwamila, B.L.M. (1979) *Flexural behaviour of concrete elements reinforced with sisal fibres.* M.Sc.Eng. Thesis, Fac. of Eng., Univ. of Dar-es-Salaam, Tanzania, 1–185.
51. Mawenya, A.S., and Mwamila, B.L.M. (1979) *Characteristics of sisal as a reinforcing fibre.* Report of Fac. of Eng., Univ. of Dar-es-Salaam, Tanzania, 1–22.
52. Persson, H. and Skarendahl, A. (1979) *Implementation of a building material–sisal fibre concrete roofing sheets.* Report 20/1/1979, Dept. of Building Eng., Royal Institute of Technology, Sweden, 1–12.
53. Swift, D.G., and Smith, R.B.L. (1979) The flexural strength of cement based composites using low modulus (sisal) fibres, *Composites* **10** (3), 145–8.
54. Swift, D.G., and Smith, R.B.L. (1979) Sisal-cement composites as low-cost construction materials. *Appropriate Technology* **6** (3), 6–8.
55. Mlingwa, G., and Mwamila, B.L.M. (1980) Local buildings materials—problems of implementation. Paper presented at the 8th CIB congress, Oslo, 1–14.
56. Johansson, L. (1981) *Corrugated sheets of natural fibre concrete–a proposal for standard test methods.* B. Eng. Thesis, Dept. of Building Eng., Royal Institute of Technology, Sweden, 3–36.

57. Persson, H., and Skarendahl, A. (1980) 'Sisal fibre concrete for roofing sheets and other purposes', *Appropriate Industrial Technlogy for Construction and Building Materials*, UN, New York, Report 12, 105–128.
58. Swift, D.G., and Smith, R.B.L. (1981) Sisal cement keeps cost down. *Kenya Builder* **5** (32), 10–11.
59. Racines, P.G. (1977) *Development of low cost roofing materials from sugarcane bagasse.* M. Eng. Thesis No. 1043, Asian Institute of Technology, Bangkok.
60. Racines, P.G., and Pama, R.P. (1978) 'A study of bagasse fibre—cement composites as low cost construction materials'. *Proc. Int. Conf. Mat. Constr. Dev. Count.*, Bangkok, 1, 191–206.
61. Fang, H.Y. (1978) *Random bamboo fibre reinforced concrete.* Report on Construction Materials Technology.
62. Smith, P.D., Amedoh, A., Nana-Acheampong, H., and Bailey, A.W. (1979) Bamboo fibre as reinforcing materials in concrete. *Appropriate Technology* **6** (2), 8–10.
63. Pakotiprapha, B., Pama, R.P., and Lee, S.L. (1979) A study of bamboo pulp and fibre cement composites *Housing Science* **3** (3), 167–90.
64. Pakotiprapha, B., Pama, R.P., and Lee, S.L. (1983) Analysis of a bamboo fibre-cement paste composite. *J. Ferrocement* **13** (2), 141–60.
65. Pakotiprapha, B., Pama, R.P., and Lee, S.L. (1983) Behaviour of a bamboo fibre-cement paste composites. *J. Ferrocement* **13** (3), 235–48.
66. Youssef, M.A.R. (1976) 'Date-palm mid-ribs as a substitute for steel reinforcement in structural concrete' *Proc. Int. Symp. New Horizons in Construction Materials*, Envo, Publishing Co. Inc., 1, 525–54.
67. Youssef, M.A.R. (1977) Concrete structural members reinforced with date-palm mid-ribs. *World Construction*, 96–102.
68. Chakravarty, A.C. (1968) Physical properties of some hard fibres used in cottage industry. *Jute Bulletin (India)* 381–92.
69. Mansur, M.A., and Aziz, M.A. (1981) 'Jute fibre reinforced composite building materials'. *Proc. 2nd Australian Conf. Eng. Materials*, Univ. of New South Wales, Sydney, 585–96.
70. Mansur, M.A., and Aziz, M.A. (1982) A study of jute fibre reinforced cement composites. *Int. J. Cem. comp. Lightweight Concr.* **4** (2), 75–82.
71. Elmendorf, A., Vaughan, T.W., and Sodhi, J.S. (1959) 'Embedded fibre board—a new low cost, fire resistant, weather-proof board for roofs, walls and floors developed in the USA'. *Proc. Symp. Timber and Allied Products*, New Delhi, India, 115–24.
72. Haines, C., and Rothrock, R.R. (1969) Process of curing hardboard containing wood fibre and Portland cement. US Patent Office 3, 4, 38, 853.
73. Sunden, O. (1977) Reinforced material system consisting of acid oligomer silicic acid-modified cellulose fibres as reinforcement material. Swedish Patent No. 398135, 1–10.
74. Andonian, R., Mai, Y.W., and Cotterell, B. (1979) Strength and fracture properties of cellulose fibre reinforced cement composites. *Int. J. Cem. Comp. Lightweight Concr.* **1** (4) , 151–8.
75. Coutts, R.S.P., and Campbell, M.D. (1979) Coupling agents in wood fibre reinforced cement composites. *Composites* **10** (4). 228–32.
76. Coutts, R.S.P. (1979) *Wood fibre reinforced cement composites.* Research Review, CSIRO Division of Chemical Technology, South Melbourne, Australia.
77. Campbell, M.D., and Coutts, R.S.P. (1980) Wood fibre-reinforced cement composites. *J. Mater. Sci.* **15**, 1962–70.
78. Campbell, M.D., Coutts, R.S.P., Michell, A.J., and Willis, D. (1980) Composites of cellulosic fibres with polyolefins or cement—a short review. *Ind. Eng. Chem. Pro. Res. Dev.* **19** (4), 596–601.
79. Mai, Y.W., Andonian, R., and Cotterel, B. (1980) On polypropylene-cellulose fibre-cement hybrid composites. *Advances in Composite Materials, Proc Int. Conf. Composite Materials*, Paris, **2**, 1687–99.
80. Mitchell, A.J. (1980) Composites of commercial wood pulp fibres and cement. *APPITA* **33** (6), 461–3.
81. Davies, G.W., Campbell, M.D., and Coutts, R.S.P. (1981) A SEM study of wood fibre reinforced cement composites *Holzforschung* **35**, 201–4.
82. Anon. (1982) New—a wood fibre cement building board. CSIRO, *Industrial News*, Council of Scientific and Industrial Research, Australia, 146.

83. Coutts, R.S.P., and Ridikas, V. (1982) Refined wood fibre-cement products. *APPITA* **35** (5), 395–400.
84. Coutts, R.S.P., and Kightly, P. (1982) Microstructure of autoclaved refined wood-fibre cement mortars. *J. Mater. Sci.* **17**, 1801–06.
85. Gram, H-E. (1982) *Cellulose fibres and other natural fibres as reinforcement of concrete products–an inventory study*. Swedish Cement and Concrete Research Institute, Stockholm Consultant Section, Report 8237, 1–18.
86. Coutts, R.S.P. (1983) Wood fibres in inorganic matrices, *J. Chemistry (Australia)* **50** (5), 143–48.
87. Coutts, R.S.P. (1983) Flax fibres as a reinforcement in cement mortars *Int. J. Cem. Comp. Lightweight Concr.* **5** (4), 257–262.
88. Uzomaka, O.J. (1976) Characteristics of akwara as a reinforcing fibre. *Mag. Concr. Res.* **28** (96), 162–7.
89. Lewis, G. (1978) 'Low-cost roofing materials for Zambia', *Proc. Int. Conf. Housing Problems in Developing Countries*, London, 111–21.
90. Lewis, G., and Mirihagalia, P. (1979) A low cost roofing material for developing countries. *Building and Environment* **14** (2), 131–4.
91. Lewis, G., and Mirihagalia, P. (1979) Natural vegetable fibres as reinforcement in cement sheets *Mag. Concr. Res.* **31** (107), 104–8.
92. Ayyar, T.S.R., and Mirihagalia, P. (1979) Elephant grass fibres as reinforcement in roofing sheets. *Appropriate Technology* **7** (2), 13–14.
93. Alexandra, J. (1976) *Indian production, exports and internal consumption of coir*. Coir Board, Cochin, India.
94. Lock, G.W. (1969) *Sisal*. 2nd edn., Longman, London.
95. Vermass, C.H. (1979) 'Evaluation of utilization of fibres from annual plants for the manufacture of resin or cement bonded particle boards'. *Proc. Seminar on Wood Processing Industries*. Cologne and Hannover, Federal Republic of Germany 1–12.
96. Mansur, M.A., and Aziz, M.A. (1983) Study of bamboo-mesh reinforced cement composites. *Int. J. Cem. Comp. Lightweight concr.* **5** (3), 165–71.
97. Mansur, M.A., and Aziz, M.A. (1983) 'Mechanical properties of bamboomesh reinforced cement composites'. *Proc. 4th Int. Conf. Mech. Behav. Materials*, Stockholm.
98. Bangladesh Jute Research Institute (1974) *Jute and jute products* Brochure of Agricultural and Industrial Exhibition, No. BGP 73/74-4351B-2000, Dacca, Bangladesh.
99. Montgomery, B. (1954) *The Bast Fibres and Leaf Fibres in Matthews Textile Fibres*. Ed. H.R. Mauersberger, John Wiley & Sons Inc., Chapters VII and VIII, 259–98, 392–405.
100. Kirby, R.H. (1963) *Vegetable Fibres*. Interscience Publishers Inc., New York.
101. Nagaraj, T.S. (1979) *Materials and construction technology—a short course lecture notes*. Quality Improvement Programme, Ministry of Education, India.
102. Jarman, C.G. (1977) 'Recent research and development in hard fibres', *Proc. Conf. Future of Natural Fibres*, Manchester, 99–109.
103. Arnaouti, C., and Illston, J.M. (1979) *The strength of cement reinforced with natural fibres*. 1st Report, Hatfield Polytechnic Div. of Civil Eng., 1–19.
104. Cook, D.J. (1980) Concrete and cement composites reinforced with natural fibres. *Concrete International*, 1980, Construction Press, Lancaster, 99–114.
105. Cook, D.J. (1980) 'Natural fibre reinforced concrete and cement—recent developments'. *Advances in Cement-matrix Composites, Proc. Symp. Materials Research Society*, Boston, 251–8.
106. Aziz, M.A., Paramasivam, P., and Lee, S.L. (1981) Prospects of natural fibre reinforced concretes in construction. *Int. J. Cem. Comp. Lightweight Concr.* **3** (2), 123–32.
107. Majumdar, A.J. (1975) Fibre cement and concrete—a review, *Composites* **1**, 7–16.
108. Pomeroy, C.D. (1976) An assessment of the commercial prospects for fibre—and polymer—modified concretes, *Mag. Concr. Res.* **28** (96), 121–129.
109. Castro, J., and Naaman, A.E. (1981) Cement mortar reinforced with natural fibres. *ACI Journal*, 69–78, and *J. Ferrocement* **11** (4), 285–301.
110. Krenchel, H., and Jensen, H.W., (1980) Organic reinforcing fibres for cement and concrete. *Concrete International*, 1980, Construction Press, Lancaster, 87–98.
111. Page, D.H., El-Hosseiny, F., and Winkler, K. (1971) Behaviour of single wood fibres under axial tensile strain. *Nature* **229**, 25231.

112. Siraskar, K.A., and Kumar, S. (1972) 'Fibre reinforcement for shatter proofness of concrete'. *Proc. Symp. Modern Trends in Civ. Eng.*, Univ. of Roorkee, India, 2549.
113. Swift, D.G., and Smith, R.B.L. (1980) 'Fibre reinforced concrete as an earthquake resistant construction material', *Proc. Int. Conf. Eng. Protec. Nat. Disaster*, Bangkok, 325–36.
114. Eldridge, J.J. (1974) *Properties of Building Boards* MTP Construction, Lancaster, England.
115. Close, P.O. (1966) *Sound control and thermal insulation of buildings.* Reinhold Publishing Co., New York.
116. Ramaswamy, S.D., and Aziz, M.A. (1983) 'An investigation on jute fabric as a geotextile for subgrade stabilization', *Proc. 4th REAAA Conference*, Jakarta, 3, 145–57.
117. Gram, H.E. (1982) *Methods to inhibit embrittlement of natural fibre concrete.* Swedish Cement and Concrete Research Institute, Stockholm, Consultant Section, Report 8201, 1–12.
118. Arnaouti, C., and Illston, J.M. (1980) *Tests on cement mortars reinforced with natural fibres.* 2nd Report, Hatfield Polytechnic Div. of Civil Eng., 1–24.
119. Parry, J.P.M. (1981) Development and testing of roof cladding materials made from fibre reinforced cement. *Appropriate Technology* **8** (2), 20–23.
120. Wells, R.A. (1982) Future developments in fibre reinforced cement mortar and concrete. *Composites*, 69–72.
121. Everett, A. (1981) *Materials.* Batsford Technical Pub., London, 227–231.
122. Robles-Austriaco, L., Pama, R.P., and Valls, J. (1983) Reinforcing with organic materials. *Concrete International*, American Concrete Institute, **5** (11), 22–26.

4 Bamboo Reinforcement for Cement Matrices

B.V. SUBRAHMANYAM

Abstract

Bamboo is a naturally occurring material with very good potential as reinforcement for cement matrices such as concrete, soil-cement, ferrocement and cement mortar. It could be used as bar type reinforcement or as a fibre. While the structural use of bamboo is perhaps as old as mankind, its use as reinforcement for cement matrices dates back to the 1910s.

The advantages of bamboo reinforcement—its low cost, its good mechanical strength ($\sim 370\ \mathrm{N/mm^2}$ in tension), its replenishable nature, and the lack of requirement of expensive and artificial forms of energy for its manufacture—are well known. Equally known are its drawbacks, viz., moisture susceptibility and dimensional changes, consequent loss of bond or cracking, and possible decay.

This chapter covers the available information on the different aspects of bamboo as reinforcement for cement matrices. To enable an appreciation of the behaviour of bamboo reinforcement, the physical structure, and the physical and engineering properties of bamboo are discussed. Information is presented on the use of bamboo as reinforcement for cement concrete structural members (beams, slabs, and columns), soil-cement, cement mortar, and a thin bamboo-cement composite called Bamboocrete. Cases of application have also been discussed.

Prophylactic treatment of bamboo to overcome its deficiencies and practical considerations in the field use of bamboo reinforcement are highlighted. Examples are presented to calculate the strength of bamboo-reinforced concrete elements. Aspects needing further research and an outlook to the future are presented.

4.1 Introduction

The concept of reinforcing cement matrices to overcome their brittle characteristics has revolutionized the construction field, making cement composites the foremost among the construction materials of the twentieth century. In its many forms as rod, mesh, and fibre, steel has established itself as a leading reinforcing material for cement composites. Unfortunately, acute shortages of steel have been experienced in several parts of the world from time to time. Also, in most of the developing countries, steel continues to be costly, scarce, and often an imported item. Asbestos fibre is another popular reinforcing material for cement sheets, pipes, and boards. The health hazards associated with the use of asbestos fibre are now well known, and asbestos fibres are expected to be withdrawn from use in the near future. On the other hand, the housing needs particularly of the developing countries, are increasing at phenomenal rates due to increasing population and urbanization associated with industrial development. Thus, there exists a great need for effective, low cost, and alternative reinforcements for cement composites. Bamboo is one of the alternative materials with a very good potential for reinforcing cement matrices, especially in the developing nations. Most of these countries are located in the tropical or subtropical belt, which is conducive to bamboo cultivation. Even if the conventional reinforcing materials are not scarce, bamboo can be advantageously used as a reinforcement in semi-permanent constructions such as housing for the economically weaker sections of the community, disaster relief structures, military structures, and a host of secondary and non-load-bearing structures, where shorter life or less stringent service requirements would be sufficient and appropriate.

Bamboo is one of the replenishable, low-cost, and low energy-intensive construction materials known to mankind since time immemorial. It is a common construction material in hut-type dwellings in the developing countries, and is also a well-established source of paper pulp. However, the superior strength and mechanical properties of bamboo, and its potential for use as a reinforcement for cement matrices were recognized only at the beginning of this century. In 1914, Chu[1] made a pioneering investigation on the use of bamboo to reinforce concrete at the Massachusetts Institute of Technology. This was followed by several field applications in China around the year 1919. Subsequently, with increasing industrial production and easy availability of steel reinforcement, interest in bamboo reinforcement waned. In 1936 there was renewed interest in bamboo reinforcement, as evident from the studies of Datta[2,3] in Germany. During the Second World War, the American and the Japanese armed forces are known to have used bamboo reinforcement in emergency military structures, thereby overcoming critical shortages of steel reinforcement. Several investigations have since been made, notably in India, the United States, Philippines, and Thailand, to understand the advantages and the limitations of bamboo as a reinforcement for cement matrices. Considering the importance of bamboo as a structural material for the developing countries, the United Nations published a report[4] in 1972, devoting a section to concrete reinforced with bamboo. This has helped to further stimulate the interest in the use of bamboo to reinforce cement matrices.

4.2 Bamboo and its characteristics

4.2.1 *Geographical distribution*

Bamboo occurs in tropical, subtropical, and even temperate regions of the world, wherever suitable ecological factors exist. Accordingly it is found in the belt extending from India to Japan (including China and south-east Asia), in Africa and Australia, and in the region extending from southern United States to Argentina and Chile. It is not found in colder temperate regions such as Canada, Europe, and the U.S.S.R. Bamboo thrives in monsoon forests and prefers well-drained sites which are not waterlogged. It dwindles into undershrubs and grasses in temperate regions[4–6]. It occurs in a variety of soils which are neither too acidic nor too alkaline. No data appear to be available on the world-wide production of bamboo. However, in India alone a total area of 0.8 million hectares is estimated to be under cultivation, yielding about 20 GN of bamboo annually[5].

4.2.2 *Botany of bamboo*

Bamboos are perennial, grasslike, woody plants belonging to the class Gramineae, and are subdivided into four families and an estimated fifty genera. Among these, bamboos of only a few genera, particularly, of *Arundinaria*, *Bambusa*, *Cephalostachyum*, *Dendrocalamus*, *Gigantochloa*, *Melocanna*, *Phyllostachys*, *Schizostachyum*, *Guadua*, and *Chusquea*, appear to possess usefulness and versatility for structural applications[4,5]. Over 1250 individual species of bamboo have been identified so far[7].

Bamboo grows in the form of a clump (Fig. 4.1), with individual culms (pointed stems) growing from a prostrate, subterranean, root-bearing rhizome. The growth of a new bamboo shoot takes place rapidly within a few months of the first growing season. Some species have been reported to grow as much as 90 cm in a single day[8]. After attaining the maximum growth, the process of maturation (lignification) of the cells starts, with concomitant increase in rigidity. The culms reach their maturity after about 3 to 5 years, depending on the species. Culms may bear secondary branches from the nodes which are at a considerable height from the base. In course of time the individual culms mature and die, while the rhizome continues to grow and regenerate new culms.

The propagation of bamboo has hitherto been mostly by natural regeneration in forests. It normally takes a period of six to twelve years for the seedlings to develop into clumps. In recent years, large-scale plantations have been established, particularly in India and Japan, by propagating bamboos by seeding, by planting nursery-raised seedlings, or by planting vegetative offsets[4]. On an average, the annual yield of bamboo per hectare ranges from 10 to 70 kN, while the yields of some species are reported to be as high as 300 kN.

Figure 4.1 A bamboo grove at the Forest Research Institute, Dehra Dun, India. Species, *Dendrocalamus giganteus.* (Courtesy of Professor Dinesh Mohan, Director, Central Building Research Institute, Roorkee)

4.2.3 *Physical structure*

The culms of bamboo are cylindrical, with diameters ranging from 2 cm to 30 cm, and lengths ranging from 3 m to 35 m. They are generally hollow, and rarely solid. The culm is divided at intervals by raised nodes, from where branches arise. The cavities of adjacent internodes are totally separated at the node by a transverse diaphragm (Fig. 4.2). The physical dimensions of the bamboo culm such as its length, diameter, and wall thickness are dependent on the species and the maturity of the culm. The diameter of the culm also decreases from the basal end to the distal end.

The exterior and interior surfaces of the culm are covered by hard waxy cuticles which resist loss of water from the culm. The culm is built up of parenchymal cells (where nutrients are stored), and vascular bundles, comprising vessels, thick-walled fibres, and sieve tubes[5]. The vascular bundles progressively become smaller in size, and denser, from the inner to

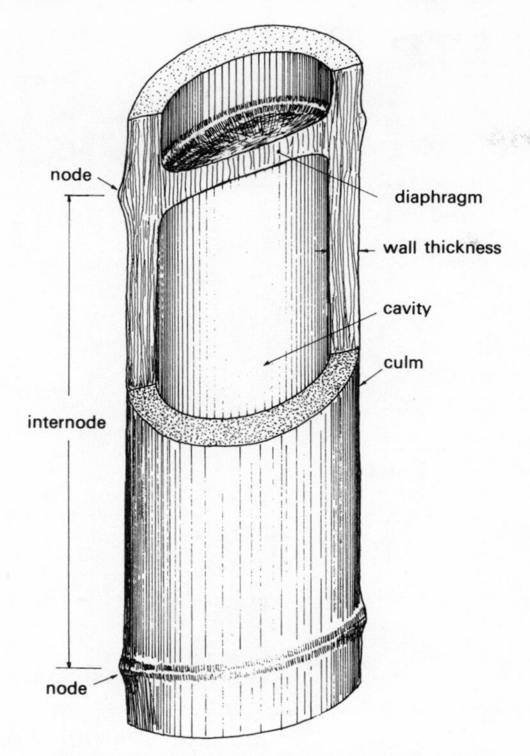

Figure 4.2 General features of a bamboo culm

the outer surface of the culm. The water movement takes place through the vessels, while the fibres are primarily responsible for the strength of bamboo[5].

The fibre content is greater at the periphery (fibre content = 40%–60%) than on the inside (fibre content = 15%–30%), where parenchymal cells predominate. The fibres are oriented parallel to the culm axis, and have an average aspect ratio (length/diameter ratio) of about 100. In the internodes, the vessels are oriented along the axis of the culm. However, at the nodes they are interconnected, and go into the diaphragm and the branches. The meristematic tissue, which is responsible for the formation of new cells and for the elongation of the internodes, is located just above the node. As a result, the node or its upper portion is generally the weakest part of the culm.

4.2.4 *Physical properties*

4.2.4.1 *Bamboo-water interaction.* Bamboo absorbs moisture, or gives it up, depending on the environment. The interaction of bamboo with water resulting in dimensional changes is of prime importance, in view of the

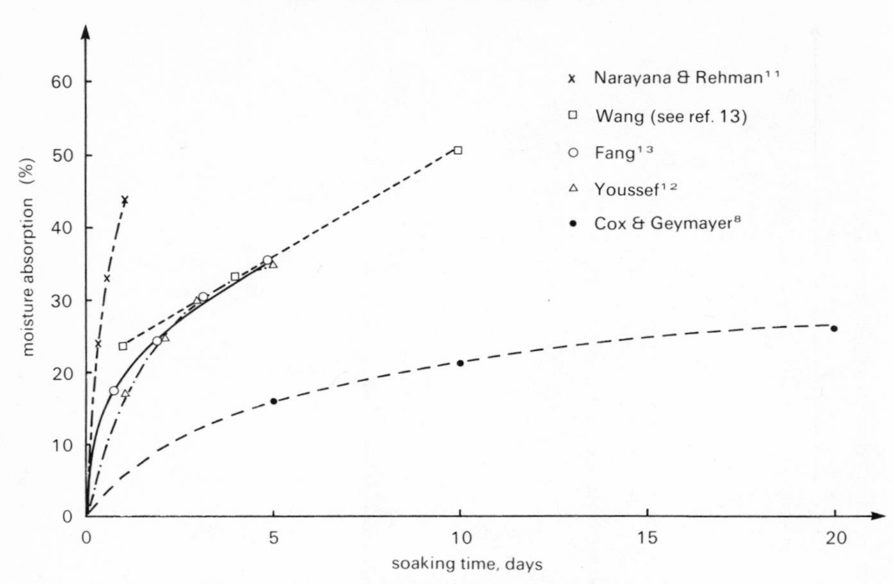

Figure 4.3 Moisture absorption of untreated bamboo with time

limitation that it imposes on the use of bamboo as a reinforcement for cement matrices. The interaction itself depends on the initial moisture content of the bamboo, among other things. The moisture content of bamboo culms decreases as the age increases; at any given age, it also decreases from the basal to the distal end of the culm. However, greater variations in the moisture content occur with changes in seasons, the amount being significantly lower in summer than in the rainy season. Accordingly, it has been suggested[9] that bamboo should be harvested from late summer to mid-autumn, as the natural moisture content is lower, and consequently the dimensional changes of bamboo are minimal. Following the harvest and during the seasoning period, an exchange of moisture takes place between bamboo and the surroundings till a state of balance, corresponding to the equilibrium moisture content, is reached. The seasoning period is typically of the order of 6 weeks.

Bamboo tends to absorb large quantities of water when soaked. Over a fairly long period, it may absorb well over 100% (by its dry weight) of water. Water absorption of even 300% has been reported by Mehra *et al.*[10] The rate of moisture absorption is quite high initially, with a large proportion of the absorption taking place within the first few days (Fig. 4.3). Subsequently, the absorption rate falls rapidly.

The absorption of moisture by bamboo is accompanied by swelling. Swelling increases as the moisture absorption continues until a limit called fibre saturation point is reached (Fig. 4.4). Beyond this limit, the change in dimensions is practically insignificant. Limited test results[8,15] indicate that the

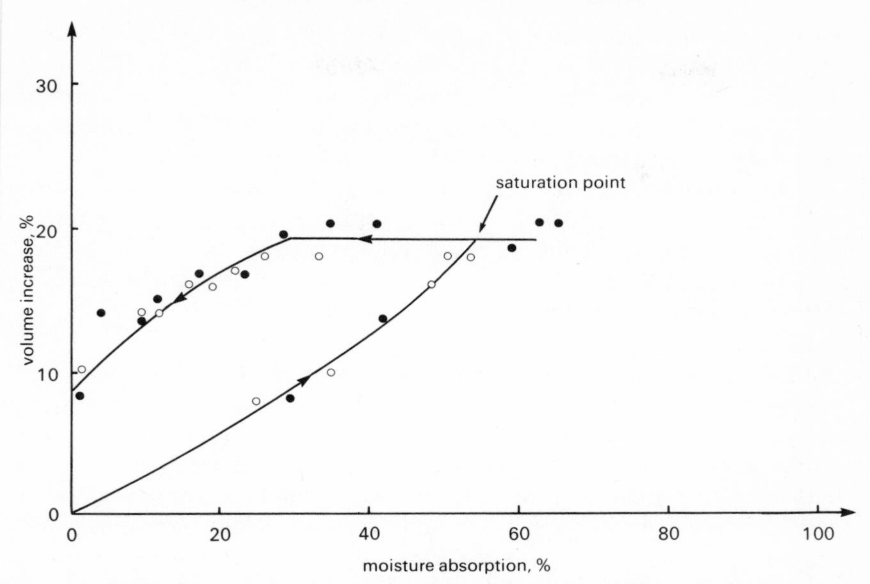

Figure 4.4 Volume change of bamboo with variations in moisture (adapted from Mehra *et al.*[14])

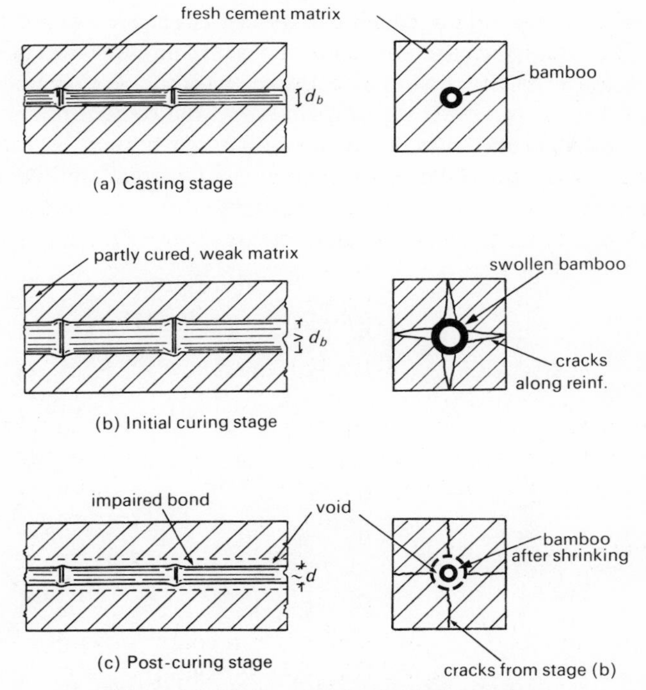

Figure 4.5 Bamboo–cement matrix interaction

expansion in the longitudinal direction is significantly low, at about 0.012 to 0.05%, in comparison to the diametrical expansion which is about 2 to 5%. This is due to the orientation of the fibres along the longitudinal direction. In general, the swelling of bamboo appears to be anisotropic, with different expansions in the longitudinal, radial, and tangential directions.

When soaked bamboo dries, the rate of shrinkage is very low or almost nil till the moisture content falls below a certain value (Fig. 4.4). Below this moisture content, the shrinkage rate is relatively high. Again, the shrinkage along the longitudinal direction is significantly smaller than that in the radial or the tangential directions.

4.2.4.2 *Mechanism of bamboo–cement matrix interaction.* It is possible to understand the interaction of bamboo reinforcement with cement matrix in the light of the bamboo-water interaction already discussed. During the casting stage and the initial curing period, the cement matrix contains relatively larger quantities of moisture which could be absorbed by the bamboo. This results in the swelling of the bamboo reinforcement and in the exertion of bursting pressures on the cement matrix, which at this stage is weak and has low strength (Fig. 4.5). Thus, cracks along the length of the reinforcement result. This also results ultimately in a reduced capacity for the member to resist horizontal shear stresses.

After the curing and the hardening of the cement matrix, when almost all of the water is used up, the bamboo begins to lose the already absorbed moisture and shrinks. This results in a void around the bamboo reinforcement and in a seriously impaired bond between bamboo and the cement matrix.

Thus the swelling and shrinking of bamboo in cement matrix poses a serious limitation to its use as an effective reinforcement. Different methods have so far been adopted to reduce the moisture movement, to and from bamboo reinforcement. These methods include one or more of the following.

(i) Use of seasoned culms so that the absorption or loss of moisture is reduced somewhat. This method, at its best, reduces the severity of cracking and of shrinkage, only slightly.
(ii) Use of pre-soaked culms, which eliminates the swelling of bamboo and the attendant cracking of the cement matrix. Loss of bond, due to subsequent shrinkage of bamboo, could still occur. Pre-soaking for a period of 7 to 10 days should normally suffice (Fig. 4.3).
(iii) Saturating seasoned culms with a non-evaporable liquid, or coating culms with a water-resistant material. A variety of materials such as varnish, asphalt emulsion[6], paints[14], rosin-alcohol mixture[14], paraffin-rosin-linseed oil mixture[16], and bitumen-kerosene mixture[11] have been tried with varying degrees of success (see Fig. 4.6) to reduce moisture absorption. However, proper care needs to be exercised in their use so that too smooth a surface with poor bond does not result. These materials must also be compatible with the cement matrix. They should

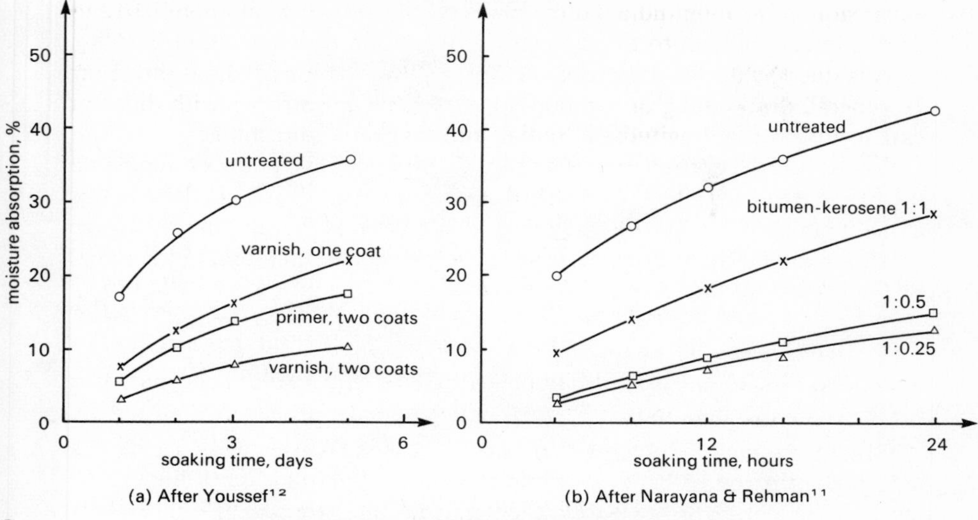

Figure 4.6 Effectiveness of some waterproofing treatments

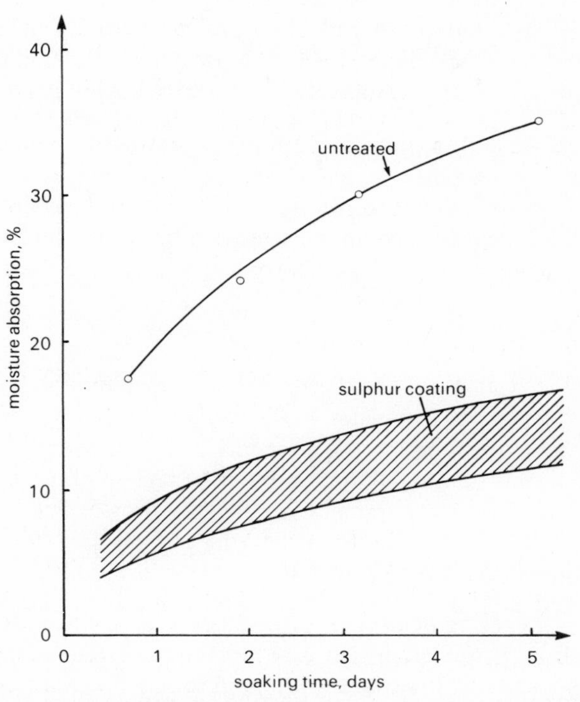

Figure 4.7 Moisture absorption of sulphur-coated bamboo (adapted from Fang and Fay[13])

not also totally prevent moisture transfer, as this could aggravate the decay of bamboo.

(iv) Yet another method is to coat seasoned bamboo culms with epoxy or polyester resin. Sand is then sprinkled on the fresh resin coating to obtain a water-resistant and rough surface[8,15,17]. With epoxy treatment, the moisture absorption has been reported[17] to be as low as 0.41%. This appears to be a very effective method; but it is also expensive, due to the high cost of the resins used.

(v) Coating bamboo with inexpensive sulphur also appears to be fairly effective (Fig. 4.7). Fang *et al.*[13] have evolved a method which involves roughening of the bamboo surface by sand blasting, wrapping a thin wire around the culm, soaking bamboo in molten sulphur at 145°C for about 1 hour, and sprinkling sand on the sulphur coating, which is then left to air-dry.

Other techniques adopted hitherto, to obviate the detrimental effects of the swelling-shrinking of bamboo reinforcement, include the use of high early-strength cement. In this case the cement matrix gains strength quickly and can resist the expansive forces due to the swelling of bamboo better. Use of high-grade concretes, in which case the matrix is stronger and the available water is lower (due to lower water to cement ratio), has also proved to be another effective method of preventing serious cracking. Adoption of shear connectors and anchorages of some sort, is yet another method which can be adopted to compensate for the poor bond between bamboo and the cement matrix. Use of bamboo splints rather than whole bamboos, and the use of proper covers and spacings are also of considerable help, to reduce the severity of cracking due to bamboo-cement matrix interaction.

4.2.4.3 *Specific gravity.* The specific gravity of bamboo increases which its age till maturity, and depends also on the species and the moisture content. Also, the outer part of bamboo is somewhat heavier than the inner part. The average specific gravity ranges from 0.3 to 0.8. The specific gravity of mature culms will be between 0.5 and 0.8[18,19].

4.2.4.4 *Thermal expansion.* The thermal expansion of bamboo is anisotropic. Its coefficient of thermal expansion in the transverse direction is significantly higher than that in the longitudinal direction. From the available information (see Table 4.1), it appears that bamboo has a coefficient of thermal expansion of 26×10^{-6} to 58×10^{-6} units/°C in the transverse direction, but only 1.4×10^{-6} to 5.2×10^{-6} units/°C in the longitudinal direction. While these values are comparable with those for wood they are considerably different from the coefficient of thermal expansion for cement concrete. In the longitudinal direction, the expansion of bamboo is less than that of concrete, while in the transverse direction, the thermal expansion of bamboo could be several times that of concrete. Since the diurnal change in temperature could

Table 4.1 Thermal expansion of bamboo

S. No	*Reference*	*Coefficient of thermal expansion ($\times 10^{-6}$ units/°C)*	
		Across fibres	*Parallel to fibres*
1	Mehra *et al.*[14]	26.1	—
2	Geymayer and Cox[15]	37.3 to 57.4	1.4 to 5.2

Note: Coefficients of thermal expansion for
(i) Concrete $= 10$ to 14×10^{-6}/°C
(ii) Steel $= 13$ to 22×10^{-6}/°C
(iii) Timber across fibres $= 32$ to 61×10^{-6}/°C
(iv) Timber parallel to fibres $= 3$ to 9×10^{-6}/°C

be as much as 10°C, the differential thermal expansion of bamboo could contribute to cracking of the concrete cover and to loss of bond[8,10,14]. It is possible to control cracking due to thermal expansion, by adopting adequate cover to the reinforcement. Mehra *et al.*[14] have developed equations to determine the minimum cover that could prevent cracking due to thermal expansion.

4.2.4.5 *Durability and resistance to fire.* Bamboo is vulnerable to attack by insects such as borers and termites, and rot fungus. Untreated bamboo in contact with ground has a maximum life of 2 years, while bamboo under cover and not in contact with ground may last from two to five years[5]. Even though bamboo reinforcement is encased inside concrete, it may still decay due to the presence of moisture, and due to insects and organisms which could attack it through the structural cracks in concrete.

Several techniques[4,5,12] are available for the chemical preservation of bamboo. However, one method which has proved to be very effective is the ASCU method[20], whereby bamboo could be protected for at least 35 years. In this method, developed by the Forest Research Institute, Dehradun, India, bamboo is treated with a solution of arsenic pentoxide, copper sulphate, and sodium dichromate.

Not much work has been done on the fire protection of bamboo. A reasonably cheap fire-resistant composition[5,4] comprises ammonium phosphate, boric acid, copper sulphate, zinc chloride, and sodium dichromate.

4.2.5 *Engineering properties*

4.2.5.1 *Factors influencing strength.* The strength of bamboo has been investigated by several research workers around the world. It is influenced by various factors, such as the species, soil and climatic conditions, harvesting age, moisture content in the samples, location of the sample with respect to the length of the culm, presence or absence of nodes in the test specimen, and

In general, the strength of bamboo is less at the nodes than in the internodes; it also decreases somewhat from the basal to the distal end of the culm.

4.2.5.2 *Behaviour in tension.* Bamboo has been reported[8,22] to possess a tensile strength as high as 370 N/mm^2, which is comparable to the strengths of reinforcing steels. This makes bamboo reinforcement an attractive alternative to metallic reinforcements. However, the strength of bamboo is subject to great variance[11]. Table 4.2 gives typical strengths and elastic moduli of bamboo as reported from different parts of the world. The tensile strength of bamboo is found to vary from as low a value as 76 N/mm^2 to a high value of 350 N/mm^2. This variation is understandable considering the influence of factors such as species, age and moisture content.

Bamboo exhibits an almost linear stress–strain behaviour in tension and fails in a brittle manner[8,11,12]. Narayana and Rehman[11] observed that the stress–strain behaviour in tension remains linear even under cyclic loads. Typical stress–strain curves are shown in Fig. 4.9*a* which demonstrates the influence of moisture content. In this case, unseasoned green bamboo containing about 39 percent moisture possessed significantly lower tensile strength and modulus of elasticity. The failure strain of bamboo was found to vary from 5×10^{-3} units[8,11] to 12×10^{-3} units[12]. As mentioned earlier, the influence of moisture on the tensile strength, in particular, and all the mechanical properties, in general, needs to be investigated more thoroughly.

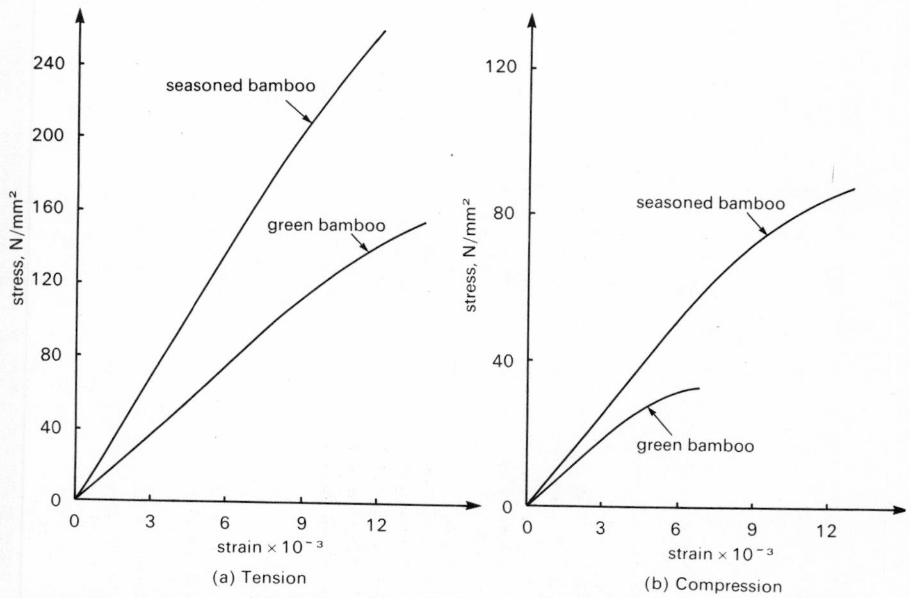

Figure 4.9 Stress–strain characteristics of bamboo in axial tension and axial compression (after Youssef[12])

Table 4.2 Strength of bamboo

(1)	(2)	(3)	(4)	(5)	(6)	(7)	(8)	(9)
S. No.	(a) *Author* (b) *Species* (c) *Location*	(d) *Age (years)* (e) *Moisture content (%)* (f) *Splint or full culm*	*Tensile*		*Flexural*		*Compressive*	
			Strength	*Modulus*	*Strength*	*Modulus*	*Strength*	*Modulus*
—	—	—	*N/mm²*	*kN/mm²*	*N/mm²*	*kN/mm²*	*N/mm²*	*kN/mm²*
1	(a) Narayana *et al*[11] (b) *Dendrocalamus strictus* (c) Roorkee, India	(d) — (e) — (f) —	102.5	13.3 to 24.6	70.8 to 134.1	—	—	—
2	(a) Limaye[4] (b) *Dendrocalamus strictus* (c) Dehradun, India	(d) 1–8 (e) Green (98.9) (f) Full	—	—	101.4	17.1	36.5	—
3	(a) Limaye[4] (b) *Dendrocalamus strictus* (c) Dehradun, India	(d) 1–8 (e) Kiln-dry (f) Full	—	—	163.3	22.1	63.4	—
4	(a) Espinosa[4] (b) *Bambusa spinosa* (c) Philippines	(d) — (e) Air-dry (f) Splint	—	—	115.8	—	52.5	—
5	(a) Mehra *et al*[14] (b) *Dendrocalamus strictus* (c) Delhi, India	(d) — (e) — (f) —	89.9	16.6	—	—	—	—
6	(a) Datta[2] (b) - - - (c) West Germany	(d) — (e) — (f) —	172.8	17.3	68.1 to 245.4	—	72.6	—
7	(a) Cox *et al.*[8] (b) *Arundinaria tecta* (c) USA	(d) — (e) 9.9 to 18.7 (f) Full	76.5 to 172.7	14.4 to 27.7	—	—	—	—
8	(a) Ali *et al.*[22] (b) *Thyrosostchya* Olivery Gamble (c) Thailand	(d) — (e) — (f) Splint	217.0 to 354.7	21.7 to 27.2	—	—	—	—
9	(a) Fang[9] (b) *Arundinaria tecta* (c) U.S.A.	(d) — (e) — (f) Full	—	—	—	—	23.2 to 27.3	7.95 to 9.50

10	(a) Cook *et al.*[23]	(d)	—	82.8*	9.65*	—	—	55.3	19.35
	(b) *Thyrosostachys* Olivery Gamble	(e) Seasoned		109.3**					
	(c) Thailand	(f)	—						
11	(a) Wu[18]	(d)	3–6						
		(e) Seasoned (~ 14%)							
		(f)	—						
	(b) i. *Phyllostachys pubescens* Mazel			—	—	156.2 to 171.9	12.1 to 13.5	72.0 to 82.5	—
	(b) ii. *Phyllostachys makinoi* Hayata		-do-	—	—	185.3 to 205.9	16.3 to 17.2	84.8 to 92.7	—
	(b) iii. *Bambusa dolichocalada* Hay		-do-	—	—	199.1 to 214.0	14.1 to 17.4	79.1 to 91.2	—
	(b) iv. *Dendrocalamus latiflorus* Munro		-do-	—	—	128.1 to 140.8	9.3 to 10.7	60.2 to 68.0	—
	(b) v. *Bambusa stenostachya* Hackel		-do-	—	—	136.6 to 140.2	10.1 to 10.5	71.1 to 73.1	—
	vi. *Bambusa oldhamii* Munro		-do-	—	—	152.9 to 166.9	12.2 to 13.9	68.2 to 69.7	—
	(c) Taiwan								
12	(a) Youssef[12]	(d) > 4		114.8*	11.3*	—	—	29.0*	4.6*
	(b) *Arundinaria gigantea*	(e) Green (39%)		161.4**	12.4**			31.7**	5.8**
	(c) Egypt	(f) Full							
		(e) Green (39%)		—	—	59.1*	6.5*	—	—
		(f) Splint				67.0**	8.1**		
		(e) Seasoned		188.4*	18.6*	—	—	87.3*	6.7*
		(f) Full		256.0**	22.2**			91.6**	9.6**
		(e) Seasoned		—	—	57.0*	15.1*	—	—
		(f) Splint				206.0**	18.6**		

Note: *At node
**In internode

There is also a need for the standardization of the tensile test specimens, as different types of test specimens have been used by different investigators. Some have preferred full bamboo rounds, while others have adopted specially-shaped bamboo splints.

The modulus of elasticity of bamboo in tension has been found to vary between 9.7 kN/mm^2 and 27.7 kN/mm^2. It has been found that bamboo has a relatively poor shear strength. It has therefore been recommended[4] that the modulus of rupture of bamboo be considered as the maximum useable tensile strength. The modulus of rupture has been reported to vary from 70.8 N/mm^2 to 245 N/mm^2, depending on the species and other factors. Again, detrimental effects due to increasing moisture content have been reported[12]. The flexural modulus of elasticity varies between 6.5 kN/mm^2 and 22.1 kN/mm^2.

Cox and Geymayer[8] found that the Poisson's ratio of bamboo ranged between 0.250 and 0.409 with an average value of 0.317. No significant difference was observed between green and seasoned bamboo.

4.2.5.3 *Behaviour in compression.* Bamboo possesses a significantly lower strength in compression than in tension. Compressive strengths ranging from 29 N/mm^2 to 88.7 N/mm^2 have been reported (Table 4.2). The stress–strain behaviour (Fig. 4.9*b*) of bamboo in compression is slightly non-linear[9,12]. The initial modulus of elasticity is of the order of 4.6 kN/mm^2 to 19.4 kN/mm^2. The strength and the modulus of elasticity decrease with increasing moisture content (Figs. 4.8, 4.9*b*). The strain at failure has been reported to vary from 3.2×10^{-3} units[9] to 12×10^{-3} units[12].

4.2.5.4 *Creep and fatigue behaviour.* When subjected to a state of stress, the cell walls of bamboo react in a time-dependent manner. This behaviour is analogous to that of wood[21]. Over long periods, wood can sustain only about 60% of its short-term strength, without failure occurring (Fig. 4.10). The limited test results of Cox and Geymayer[8] on bamboo are in qualitative agreement with this. They observed that bamboo tensioned to 27.6 N/mm^2 did not fail over a period of 1 year, while failure occurred within 10 minutes to 188 days under a stress of 56.9 N/mm^2. The short-term strength of bamboo in this particular case ranged from 80.4 N/mm^2 to 119.2 N/mm^2. (There was, however, one exceptionally low value of 47.6 N/mm^2.) In the absence of more reliable data and guidelines, the sustained strength of bamboo may be taken for design purposes to be about 50% of its short-term strength.

Cox and Geymayer also observed that the creep coefficient (ratio of creep strain to elastic strain) was about 0.4 for both the stress levels which they considered. The creep deformations were of the order of 17.4×10^{-6} to 20.3×10^{-6} units per N/mm^2. These tests also indicated that shrinkage or expansion strains are induced due to variations in the temperature and the relative humidity of the surroundings.

No information is available on the sustained strength of bamboo under

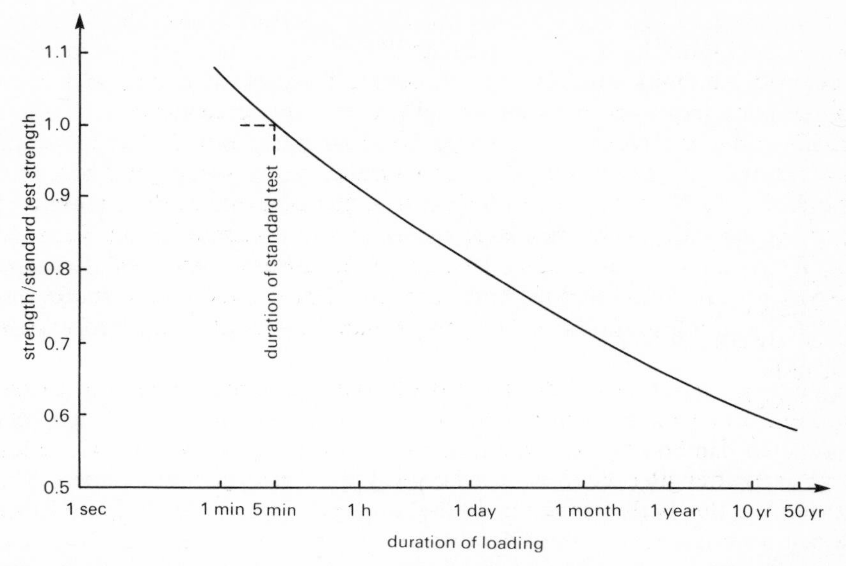

Figure 4.10 Influence of the duration of loading on the strength of wood (after Panshin and de Zeeuw[21])

either compression or flexure. So also the influence of cyclic loads on the strength of bamboo is not yet known.

4.2.5.5 *Bond between bamboo and cement matrix.* For the effective realization of the potential of bamboo as a reinforcing material, good bond with the cement matrix is necessary. Several factors contribute to the rather poor bond between bamboo and the matrix. The natural surface of bamboo is smooth. Good bond cannot therefore be expected, even though the nodes with their protrusions could act as intermediate anchorages. Dimensional changes of bamboo due to moisture and temperature variations affect its bond characteristics seriously. The swelling of bamboo during the casting and curing of the cement matrix, and its subsequent shrinkage result in a void around the reinforcement. The bond strength is thus significantly reduced. So also the differential thermal expansion of bamboo with respect to the matrix could lead to cracking of the matrix and loss of bond strength.

Considerable variance is found in the bond strength of bamboo in cement concrete, due to the interaction of the several factors involved. The principal factors influencing the average bond strength include the treatment and condition of the bamboo, size of protrusions and the spacing of nodes, age and curing conditions of the concrete, and the strength and other properties of the concrete itself. Seasoned bamboos possess higher bond strengths compared to unseasoned (green) bamboos due to lower water absorption of the former, and the reduced swelling and shrinkage of bamboo[6,12]. The bond strength of green and unseasoned bamboo is highly unreliable[12]. Values between 0.29 N/mm^2 and 1.18 N/mm^2 have been reported[4,10,12,17,18] for the bond strength of seasoned bamboos without any treatment. Presence of nodes, with their protrusions, improves the bond strength. Use of split culms instead of whole bamboo did not result in improved bond strength. But due to increased surface area, the total pull-out force increases correspondingly. Thus, split bamboo is to be preferred to whole culms. The bond strength of presoaked bamboo depends on its shrinkage subsequent to concrete curing. Thus the bond strength could decrease as the age of the member increases[8]. Cox and Geymayer also found that the bond strength of presoaked, split bamboo was much lower in beams (only 0.14 N/mm^2) than the strengths obtained in pull-out tests.

Youssef[12] observed that the bond strength increases with increasing cement content and with reducing water to cement ratio, for both treated and untreated bamboo. Similarly, increase in bond strength was observed when high early strength cement was used instead of ordinary portland cement. The reason for this is the reduction in the amount of water which could cause dimensional changes, or the reduction in the duration of exposure of bamboo to such water.

It is well recognized now that an effective moistureproof treatment of some sort is necessary for bamboo reinforcement. Several different treatments have been adopted with varying degrees of success. An effective treatment is required to provide moisture resistance; but it should not impair the bond between bamboo and matrix by the formation of any smooth surface coating. Significant improvements in bond have been reported by depositing fine sand on freshly-treated bamboo. Table 4.3 lists typical bond values for bamboo with cement concrete for some of the more successful treatments. Among these, the use of resins like polyester and epoxy, even if costly, is most effective. The cheaper treatments using bitumen, varnish, and sulphur also result in bond strengths ranging from 0.8 N/mm^2 to 1.3 N/mm^2. These values are still lower than the values for plain, mild steel reinforcement (permissible average bond stress = 0.7 N/mm^2)[25]. However, they could be adequate considering that the permissible tensile stress for bamboo is much lower than that for mild steel reinforcement.

Further improvements in the bond strength appear to be possible by winding metallic wire[8,9] or coir rope[24] around the bamboo reinforcement. Use of nails as spikes along the length of bamboo has also been found[24] to be very effective.

An alternative method for transferring the stresses from the bamboo reinforcement to the surrounding concrete involves the use of either intermediate or end anchorages (shear connectors). This method was first attempted by Narayana and Rehman[11]. They tried integral shear connectors (formed by cutting notches at intervals in whole bamboos), as well as plate shear connectors (metal plates driven into whole bamboo in a direction

Table 4.3 Bond strength of bamboo with cement concrete

S. No.	Author	Treatment	Moisture absorption after treatment	Bond strength	Remarks
—	—	—	%	N/mm^2	—
(1)	Narayana and Rehman[11]	(i) Untreated	43.0	—	Seasoned, whole bamboo.
		(ii) Bitumen–kerosene (1:0.25) coating	13.0	—	-do-
(2)	Cox and Geymayer[8]	(i) Untreated	—	0.34 to 0.38	Seasoned, split bamboo.
		(ii) Wound with steel wire	—	0.52	
		(iii) Polyester–sand	—	> 0.56*; 0.81	Only two results (*tension failure)
		(iv) Epoxy–sand	—	1.13; > 1.15*	Only two results (*tension failure)
(3)	Murthy and Deshpande[24]	(i) Bitumen–sand	—	—	Considerable improvement, and success claimed.
(4)	Youssef[12]	(i) Untreated	35.0	0.56^{+} 0.68^{++}	Split bamboo ($^{+}$Cement content = 250 kg/m^3; $w/c = 0.6$
		(ii) Varnish (2 coats)	11.0	1.11^{+} 1.27^{++}	$^{++}$Cement content = 350 kg/m^3; $w/c = 0.4$)
(5)	Gupchup *et al.*[17]	(i) Untreated	37.5	0.29	
		(ii) Bitumen	9.8	0.84	
		(iii) Epoxy	0.4	1.13	
(6)	Fang and Mehta[9]	(i) Untreated	—	3.2 to 12.8 kN$^{@}$	whole bamboo
		(ii) Sulphur—sand	—	10.3 to 13.8 kN$^{@}$	($^{@}$Total pull-out force)
(7)	Cook *et al.*[23]	(i) Untreated	—	0.24 to 1.47 (average 0.35)	Seasoned bamboo

(a) Shear connectors proposed by Narayana and Rehman[11]

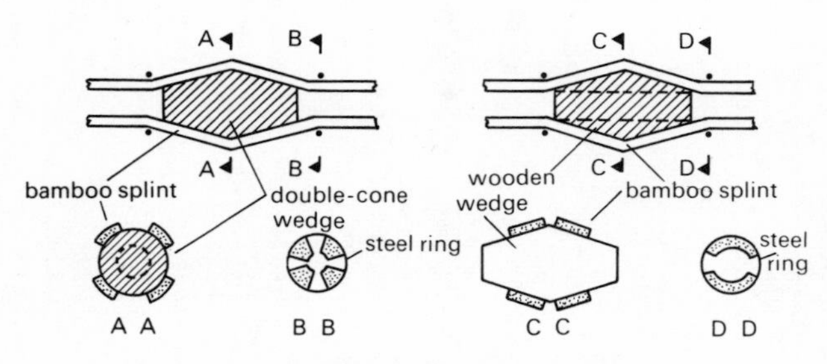

(b) Shear connectors suggested by Datye *et al.*[27]

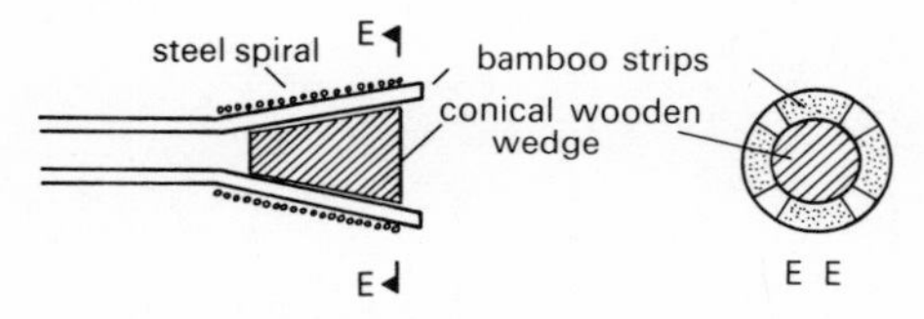

(c) End anchor (Datye *et al.*[27])

Figure 4.11 Shear transfer devices for bamboo

perpendicular to the length of bamboo) (Fig. 4.11*a*). The plate shear connectors did not perform well, from the considerations of strength and stiffness. On the other hand, the integral shear connectors could withstand a maximum shear of 6.4 N/mm^2 on the shearing area of the connector. Cox and Geymayer[8] also found the use of integral shear connectors to be an effective method, which does not require any special materials. However, this method involves considerable effort for the preparation of the reinforcement. Moreover, a major portion (almost half) of the useable bamboo reinforcement is wasted in this method.

Datye *et al.*[26,27] have recently proposed several interesting schemes (Figs. 4.11*b*, *c*) for the intermediate and end anchorages for split bamboo reinforcement in low-modulus concretes. Shear transfer capacity up to half the tensile strength of bamboo has been reported with these bulb or shear connectors. These anchorage schemes hold great promise for use in cement matrices, and when developed, they could solve the bond-anchorage problem of bamboo reinforcement in cement matrices.

Table 4.4 Bond between bamboo and other cement matrices.

S.No	*Author*	*Cement matrix*	*Bond strength*	*Remarks*
—	—	—	*N/mm²*	—
(1)	Mehra *et al.*[14]	Soil-cement		Plasticity index of
		(i) with 7% cement	0.62	soil = 8.6.
			0.48*	Cured originally for
		(ii) 10% cement	0.70	
			0.70*	one week.
		(iii) 15% cement	1.16	(*After 12 wetting-
			0.89*	drying cycles)
(2)	Kalita *et al.*[28]	Cement mortar	0.3 to 0.4	—

Very limited information is available on the bond between bamboo reinforcement and other cement matrices such as soil-cement. Mehra *et al.*[14] observed that bond between treated bamboo splint and soil-cement increased with increasing cement content (Table 4.4). Further, presoaked and painted bamboo did not lose bond significantly, even after several wetting and drying cycles. This was due to the significant shrinkage of soil-cement which compensates the shrinking of bamboo during the drying cycle. Chadda[29], however, stresses the need for proper selection of the soil. Kalita *et al.*[28] reported the bond between bamboo and cement mortar to be of the order of $0.3\,N/mm^2$ to $0.4\,N/mm^2$.

4.3 Bamboo as reinforcement for cement composites

Due to its low cost, easy availability, and reasonably high strength, bamboo can be used to reinforce cement matrices replacing conventional, relatively scarce, materials such as mild steel and galvanized steel mesh, or fibres such as asbestos. Thus, bamboo could be used to reinforce cement concrete flexural and compression members; and soil-cement elements. Meshes made of bamboo splints could be used to reinforce cement mortar to obtain thin, ferrocement-like material[30]. Like other vegetable fibres, bamboo fibre could also be used to reinforce cement concretes and mortars. Each of these applications has so far been developed to different degrees. The considerations governing these applications are also different, and need separate discussion.

4.3.1 *Bamboo reinforced conrete (BRC) structural elements*

4.3.1.1 *BRC beams.* The possible use of bamboo as tensile reinforcement in flexural members such as beams and slabs has long been recognized[1–3]. Soon it was discovered that the greatest hurdle in the exploitation of the potential of bamboo reinforcement is the dimensional change due to moisture variations.

Unless effectively treated, bamboo reinforcement tends to absorb moisture from fresh concrete, and swells, leading to severe cracking of the concrete even before any load is applied. Subsequently, the bamboo shrinks as moisture is lost, losing bond strength which is vital for its effective function as reinforcement. This very important aspect has been well recognized, and several prophylactic treatments and procedures have been developed. These treatments are effective to varying degrees, and involve some additional cost and effort. Most of the investigators have tended to avoid the more effective treatments as being costly and involved, and have preferred simple treatments of intermediate and uncertain effectiveness. For example, for expedient, military construction, Cox and Geymayer[8] consider simple treatments such as presoaking of seasoned bamboo prior to the casting of concrete. While such a treatment could, at its best, totally eliminate the swelling cracks in concrete, adequate bond between bamboo and concrete cannot be assured. The variance in the treatments and their effectiveness have naturally resulted in wide variations in the structural responses observed and in underrating the very usefulness and reliability of bamboo as a tensile reinforcement. Figure 4.12 shows the structural responses of beams with bamboo reinforcement, treated using two different methods. Due to more effective treatment of the reinforcement, Beam *A* which contained lesser reinforcement ($p_b = 3.53\%$) was stiffer than Beams *B* and *C*, and failed at significantly higher loads. Of beams *B* and *C*, both containing 4.53% reinforcement, the former was more effective

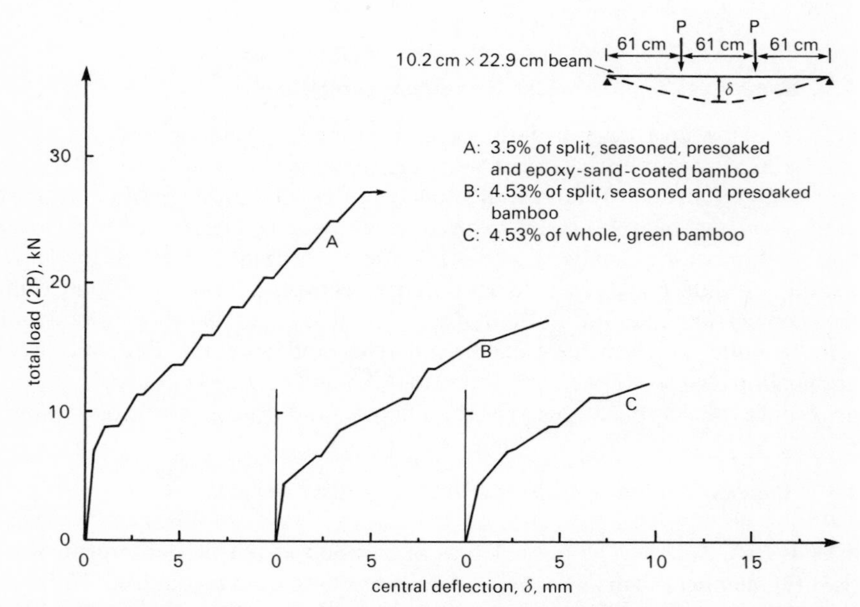

Figure 4.12 Deflection behaviour of beams reinforced with bamboo (adapted from Cox and Geymayer[8])

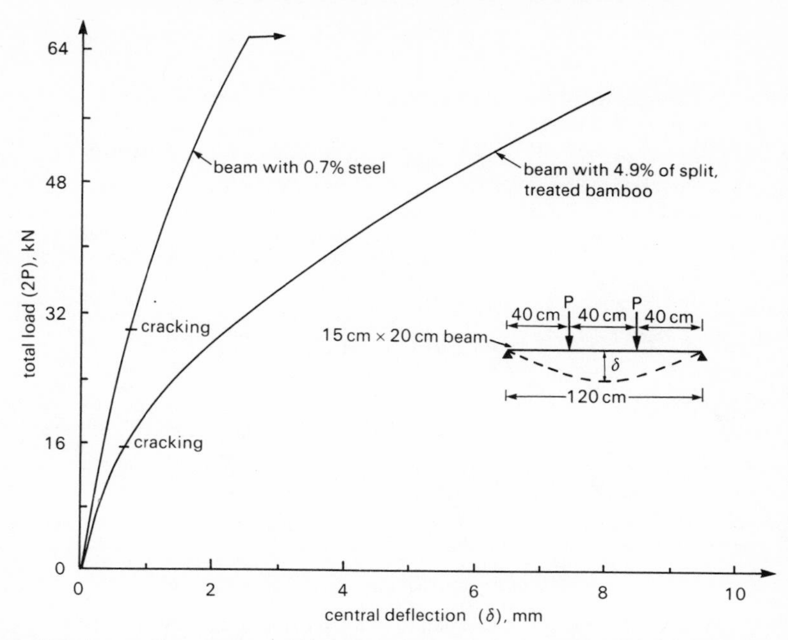

Figure 4.13 Comparison of the deformation behaviour of beams with steel and bamboo reinforcements (after Youssef[12])

since the bamboo was used in the form of splints and after seasoning and presoaking.

The structural response of bamboo-reinforced members in flexure has been investigated in detail by Glenn[6], Narayana and Rehman[11], Cox and Geymayer[8,15], Youssef[12], and Gupchup *et al.*[17], among others. The flexural response of bamboo-reinforced members is very similar to that of a steel-reinforced member (Fig. 4.13). The load–deformation response till the first cracking load has been observed to be linear[6,8]. Bamboo reinforcement may not substantially change the cracking moment from that for an unreinforced element[6,15]. Indeed, a reduction in the cracking moment is possible due to the low modulus of elasticity of bamboo, and the possible cracking of concrete due to swelling of bamboo. Thus, Cox and Geymayer[8] and Youssef[12] reported reductions in the cracking moment. On the other hand, Gupchup *et al.*[17] reported an exceptional and anomalous case of improvement in cracking moment with the use of epoxy-resin-treated bamboo.

Subsequent to flexural cracking, the load–deflection response maintains a near-linear relationship, but with a reduced slope (Fig. 4.14). The slope of the deformation response depends on the quantity of the reinforcement; the greater the reinforcement, the stiffer the response. Also, the effectiveness of the bamboo to concrete bond influences the behaviour significantly. Figures 4.12 and 4.14 present the load-deflection response (due to Cox and Geymayer[8]).

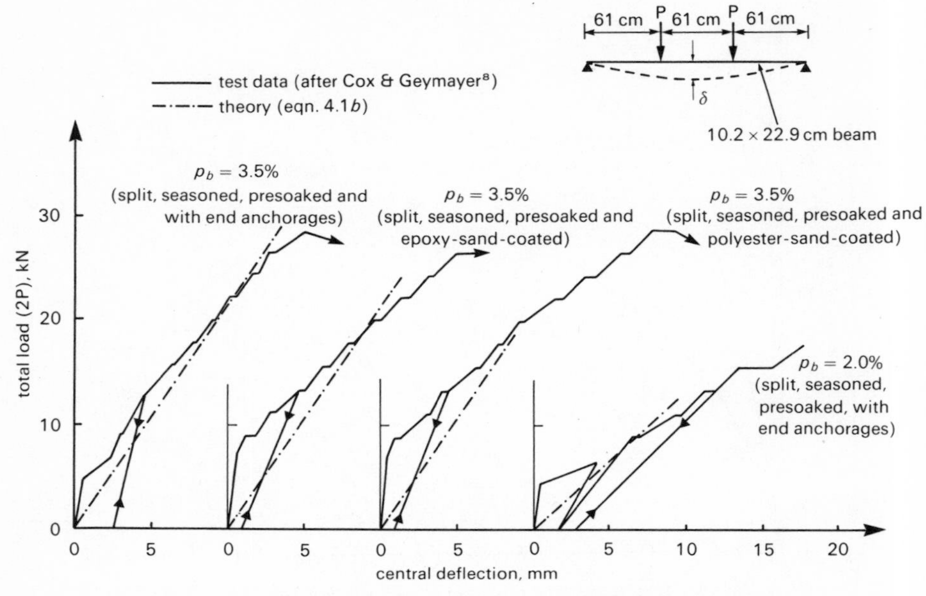

Figure 4.14 Deflection prediction for bamboo reinforced concrete beams

Beams containing 3.5% reinforcement, treated with either epoxy-sand or polyester-sand, or having effective end anchorages in the form of whole culms, exhibited lower deflections than a beam with a higher amount ($p_b = 4.53\%$) of presoaked and untreated reinforcement. This is in concurrence with the observations of Narayana and Rehman[11]. Similarly, the absence of adequate shear reinforcement would result in diagonal cracking and in additional deflection. It would obviously result in a reduced load-carrying capacity.

Bamboo-reinforced concrete members deflect considerably more than the corresponding steel reinforced members (Fig. 4.13). Currently, no proven methods are available for the estimation of the deflections, and for limiting them under service conditions. Obviously, the poor bond characteristics of the bamboo reinforcement, resulting from ineffective treatment, complicate the development of a method for estimating the deflections. Indeed, Cox and Geymayer[8] mention one of the conclusions of an earlier investigation that the deflections of bamboo-reinforced concrete beams cannot be estimated using the moment of inertia of either the cracked section or the uncracked and transformed section. However, if the bamboo reinforcement were to be given an effective treatment ensuring adequate bond it appears that the deflections under service loads could be estimated with reasonable accuracy.

The moments of inertia for a BRC beam in the uncracked and cracked stages can be obtained (see Appendix A for derivation of the equations) as

$$I_1 = \frac{bD^3}{12} \tag{4.1a}$$

$$I_{cr} = \frac{bd^3}{3}\left[\frac{k^3}{3} + \mu m(1-k)^2\right] \tag{4.1b}$$

where I_1 = gross moment of inertia for the uncracked section.
I_{cr} = cracked moment of inertia.
b = width of the member
D = total depth of the member
d = effective depth of the member
μ = reinforcement ratio
m = ratio of modulus of elasticity of the reinforcement to that of concrete.

Figure 4.14 compares the predictions using the above equations for some of the beams reported by Cox and Geymayer[8]. The split bamboo reinforcement in these beams was effectively treated using either epoxy resin-sand or polyester resin-sand, or they had end anchorages consisting of whole culms. It can be seen that under service loads, the theoretical predictions using the cracked section moment of inertia give values slightly on the conservative side. The cracked section moment of inertia can therefore be safely used to derive, on a conservative basis, the limiting span to effective depth ratios for achieving deflection control in concrete members containing effectively-treated bamboo reinforcement. Figure 4.15 presents the limiting span to effective depth ratios, for controlling the short-term deflection under service loads of BRC beams to less than span/360. (See Appendix A for derivations.) It appears that, from deflection considerations, the stress in bamboo under service loads should not exceed 20 to 30 N/mm^2, if the beam depths are to be reasonable. It is interesting to note that values of 21 to 28 N/mm^2 are recommended in Ref. 4

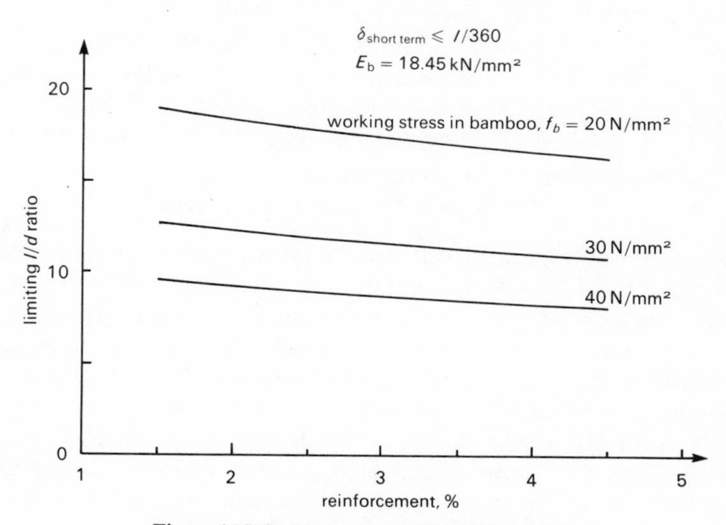

Figure 4.15 Deflection control in BRC beams

for the permissible tensile stress in bamboo reinforcement, from deflection considerations.

Due to the poor bond characteristics of bamboo reinforcement, cracks form at wide spacings and are not numerous. Typically, in a beam with 3.5% of well-bonded (polyester resin-sand treated) bamboo reinforcement, the individual crack spacings could be estimated from Ref. 8 to be 33.6 cm and 69.9 cm, at a load equal to 0.54 times the failure load. The occurrence of wide cracks under service loads has often been reported (see Ref. 8, for example). Unfortunately, no quantitative information is available so far. Such wide cracking is to be expected, in view of the large crack spacing, poor bond characteristics, and the large strains in bamboo (due to low modulus of elasticity). For steel reinforcement, the crack width equation of the following form is generally assumed[31]:

$$w = a(\varepsilon_r - \varepsilon_{cm}) \tag{4.2}$$

where w = average crack width (in mm.)
a = average crack spacing (in mm.)
ε_r = average strain in reinforcement
ε_{cm} = average strain in concrete at the level of the reinforcement.

From considerations of aesthetics and the prevention of corrosion of the reinforcement, the current practice with steel-reinforced members is to limit the average crack width to 0.3 mm, 0.2 mm, or 0.1 mm, for interior structures, for outdoor structures not exposed to severe conditions, and for structures exposed to aggressive environment, respectively.

In the absence of a better guideline, if Eq. (4.2) is assumed to be valid also for bamboo-reinforced members, and if the typical values

$a = 518$ mm (as estimated for Beam No. 19 of Ref. 8 with 3.5% reinforcement)

$\varepsilon_r = 1.63 \times 10^{-3}$ (corresponding to a permissible stress in bamboo of 30 N/mm^2, and a modulus of elasticity of 18.45 kN/mm^2)

and $\varepsilon_{cm} = 0$ to 0.5×10^{-3} (for poor and good bond characteristics, respectively)

are assumed, the average crack width under service load can be estimated to be of the order of 0.84 to 0.59 mm. These values are materially in excess of the currently permissible limits, albeit for steel-reinforced members. While these crack width values may perhaps not violate the aesthetic requirements, adequate durability cannot be assumed without extensive and long-term field investigations. Data on this aspect are seriously lacking.

In BRC members, even under service loads, the cracks extend very close to the compression face, indicating a high position of the neutral axis. The stresses in the bamboo reinforcement and the concrete can be calculated on the basis of an elastic analysis such as that presented in Appendix B. The permissible tensile stress in bamboo reinforcement has been recommended variously as 21 to 28 N/mm^2[4,6], 24 N/mm^2[20], 35 N/mm^2[8,15], 37 N/mm^2[17], 35 to 40 N/mm^2[12], and 40 N/mm^2[32]. As observed earlier, crack-width considerations have not been considered in arriving at these permissible stresses. Another important aspect which apparently has not been considered in arriving at the permissible stress under service loads, is the negative influence of sustained loading on the strength of bamboo reinforcement. Earlier, it was observed that the sustained strength of bamboo reinforcement could be as low as 50% of its short-term strength. The influence of sustained loading on the strength and behaviour of BRC members is therefore of great importance. No information on this aspect is available so far, and only future research can answer the related, vital questions. In the interim, it will be prudent to adopt an additional factor of safety of 2.0 to account for the influence of sustained loading on the strength of bamboo. The total factor of safety for the stress in reinforcement would thus be 2.0 times the load factor (say 1.75) = 3.5. Considering all these aspects, until results of further research are available, the permissible stress (under working loads) in bamboo reinforcement may be limited to the lower of (i) tensile strength/3.5, (ii) the stress limitation from deflection consideration (20 to 40 N/mm^2), and (iii) tensile stress at bond failure/factor of safety (1.75).

From the limited information available[8,11], it appears that repeated loading of a few cycles is not likely to have any adverse effect on the structural behaviour of BRC members. However, no information appears to exist on the fatigue behaviour of bamboo-reinforced beams.

Unlike steel-reinforced members, BRC members do not exhibit a plastic or post-yield behaviour, and no redistribution of internal forces is possible in an indeterminate member. These members fail abruptly when the bamboo reinforcement reaches its maximum strength, or when the reinforcement slips continuously through concrete (see Figs. 4.12 and 4.14). Cox and Geymayer[8] have proposed two alternate methods, one based on elastic theory and the other as a modification of the ultimate strength theory given in ACI 318 code[33]. Both of these methods have serious drawbacks in the case of BRC members. A more rigorous inelastic method is proposed in Appendix B, for the calculation of the ultimate strength of bamboo reinforced members, utilizing the strain compatibility and the equilibrium conditions, and the compressive stress–strain characteristics of concrete. For most normal cases, it appears (Appendix B) that the ultimate or failure strength M_f can be obtained using the simple equation

$$M_f = (\mu j f_{bf}) b d^2 \tag{4.3}$$

where μ = reinforcement ratio

j = lever arm factor (assumed as 0.9 to 0.95)

f_{bf} = stress at failure in bamboo (i.e. its usable

tensile strength, or the stress at which
it slips continuously due to bond failure)
b = width of the member
d = effective depth of the member.

Even though the tensile strength of bamboo has been observed to be fairly high, ranging from 76 N/mm^2 to 350 N/mm^2, from deflection consideration and considering the possible influence of sustained loading, the usable tensile strength (f_{bf}) of bamboo reinforcement will be the lower of (i) tensile strength of bamboo divided by a factor 2.0 to account for sustained load, (ii) factor of safety times the stress limitation from deflection consideration (i.e. 1.75 times 20 to 40 N/mm^2 = 35 to 70 N/mm^2), and (iii) tensile stress at bond failure. This useable tensile strength of bamboo is roughly one-seventh to one-fourth of the yield strength of mild steel. It is therefore not surprising that some investigators[12] have suggested that the quantum of bamboo reinforcement should be as much as 10 times the quantity of mild steel reinforcement, for obtaining equal moment capacities. However, it is not possible to locate effectively more than about 3–4% of bamboo reinforcement in a member. In fact, use of reinforcement in quantities greater than about 3–4% has been observed to reduce the load-carrying capacity[6,10,24]. Thus, the optimum quantity of bamboo reinforcement appears to be about 3–4%.

The possibility of using bamboo as shear reinforcement, either in the form of vertical splint reinforcements or by cranking up the longitudinal reinforcement, has been studied[4,9,12]. Both these methods have not been found to be effective, for different reasons. In the first case, adequate bond development length will not be available; while in the second case, the reinforcement could break on account of its relatively lower flexural strength. Thus, it appears that steel stirrup reinforcement would be necessary, to avoid shear failure and distress.

To make successful applications with bamboo reinforced concrete, several factors need to be additonally considered in construction. Appendix C gives some of these practical considerations.

4.3.1.2 *BRC slabs.* The structural responses of bamboo-reinforced, one-way and two-way slabs have been found[8,11,15,17,20] to be very similar to the behaviour of beams discussed above. Due to the relatively lower shear stresses in slabs, the problem of bond will not be very critical. However, Narayana and Rehman[11] observed that the use of integral shear connectors (Fig. 4.16) resulted in significantly improved load-carrying capacity and stiffness of one-way slabs, while other methods were not so effective. The load–deformation responses were also observed to be very similar to those for BRC beams (Fig. 4.12).

In their tests on two-way slabs, Cox and Geymayer[8,15] observed the formation of yield line patterns very similar to that for steel-reinforced slabs.

Figure 117 Load–deformation response of bamboo-reinforced concrete column (after Youssef)

compressive strength of columns needs to be assessed carefully as such very limited information is available for the behaviour of bamboo-reinforced columns.

Datta [illegible] reported that bamboo [illegible] sufficient quantity improved the compressive strength of weaker concrete. Youssef [illegible] observed that the load–deformation behaviour in axial compression of columns with steel [illegible] or bamboo reinforcement [illegible] was very similar (Fig. 117). He also concluded that if [illegible] the reinforcement area [illegible] be provided by bamboo [illegible] the failure load is at least equal to that for a steel-reinforced column. This conclusion does not appear to be correct. Since the contribution of [illegible] of mild steel reinforcement [illegible] the strength of the column considering the reported [illegible] strength [illegible] cannot exceed [illegible]. Thus the only observation that could be made is that even [illegible] of bamboo reinforcement did not significantly alter the axial failure load of [illegible] the column. Indeed, Cook [illegible] confirm this observation. They reported that a column with [illegible] of bamboo reinforcement [illegible] at a load not greatly lower than the failure load of an unreinforced concrete column. [illegible] however, they reported that bamboo [illegible] cases [illegible] the capacity of concrete columns [illegible] part of the column is subjected to tensile stresses. Accordingly, [illegible] eccentricity to depth ratio [illegible] no significant improvement in load was observed (Fig. [illegible]). At higher [illegible] ratios, the failure [illegible] for unreinforced columns should [illegible] theoretically [illegible] in practice [illegible] bamboo-reinforced column [illegible] fail at higher loads. Cook [illegible] also [illegible] that the [illegible]

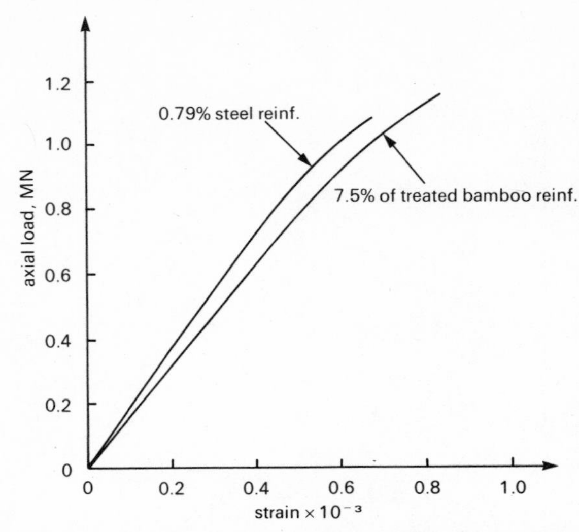

Figure 4.17 Load–deformation response of bamboo reinforced concrete columns (after Youssef[12])

compressive strength of columns needs to be assesed carefully. As such, very limited information is available on the behaviour of bamboo-reinforced columns.

Datta[2,3] reported that bamboo struts in sufficient quantity improved the compressive strength of weaker concrete. Youssef[12] observed that the load–deformation behaviour in axial compression of columns with steel (0.79%) or bamboo reinforcement (7.5%) was very similar (Fig. 4.17). He also concluded that 10 to 11 times the reinforcement area must be provided in a bamboo-reinforced column in order that the failure load is at least equal to that for a steel-reinforced column. This conclusion does not appear to be correct, since the contribution of 0.79% of mild steel reinforcement to the compressive strength of the column (considering the reported cube strength of 38 N/mm^2) cannot exceed 10%. Thus the only observation that could be made is that even 7.5% of bamboo reinforcement did not significantly alter the axial failure load of a concrete column. Indeed, Cook *et al.*[23] confirm this observation. They reported that a column with 7.28% of bamboo reinforcement failed at a load marginally lower (2.7%) than the failure load of an unreinforced concrete column. The cube strength of concrete in these columns was 34.5 N/mm^2. Further, they observed that bamboo reinforcement increases the load carrying capacity of concrete columns only when part of the column is subjected to tensile stresses. Accordingly, for eccentricity to depth (e/h) ratios up to 0.3, no significant improvement in load was observed (Fig. 4.18). At higher e/h ratios, the failure loads for unreinforced columns would be theoretically zero (a small value, in practice), while bamboo-reinforced columns would fail at higher loads. Cook *et al.* also proved that the load–

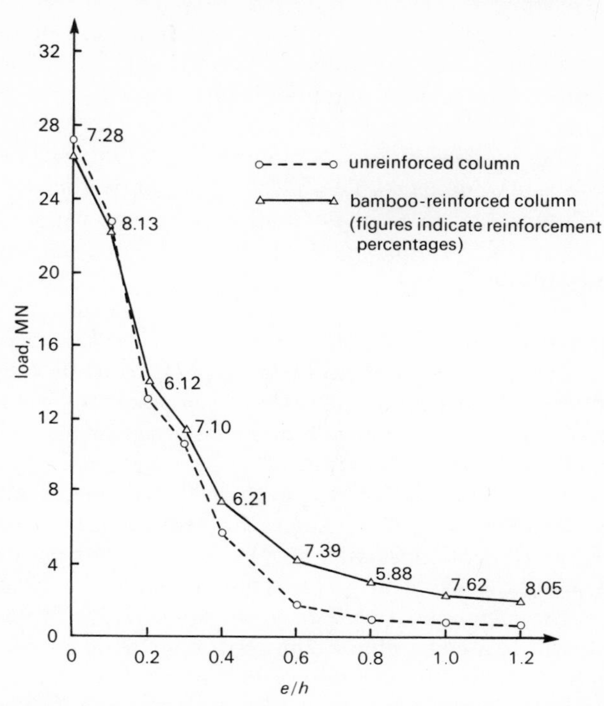

Figure 4.18 Strength of BRC columns in compression and bending (based on data from Cook *et al.*[23])

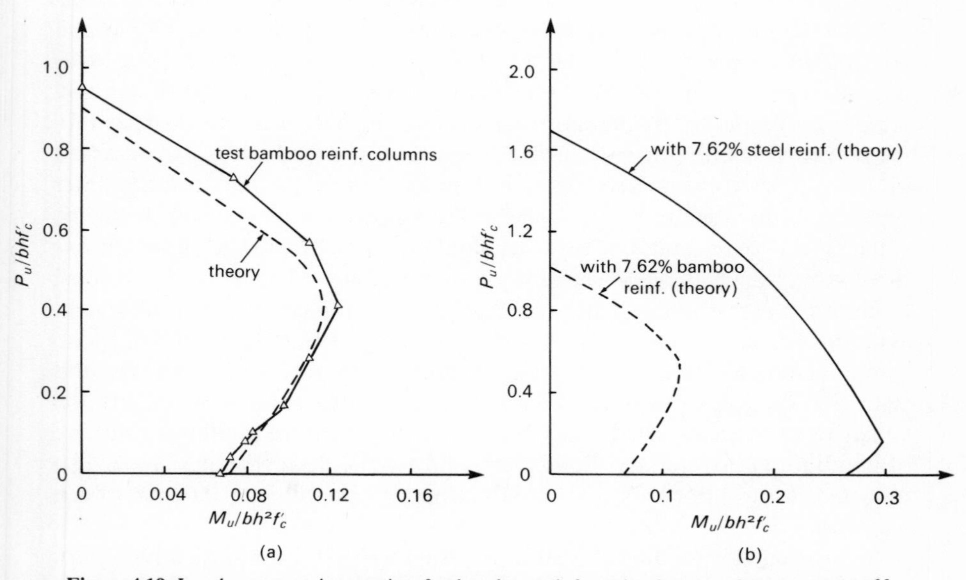

Figure 4.19 Load–moment interaction for bamboo-reinforced columns (after Cook *et al.*[23])

moment (P–M) interaction curves for bamboo-reinforced columns could be accurately predicted, talking cognisance of the characteristics of bamboo (Fig. 4.19*a*). For equal reinforcement quantities, bamboo-reinforced columns resist between 40 and 25% (for $e/h=0$ and $e/h=1.2$, respectively) of the load carrying capacities of steel reinforced columns (Fig. 4.19*b*). In spite of this, the bamboo reinforced columns would still be significantly cheaper than steel-reinforced columns[23]. The ultimate load capacity of bamboo-reinforced columns under biaxial loading could be predicted using the failure-surface method due to Bresler[34,35].

4.3.1.4 *Applications of bamboo-reinforced concrete.* The first recorded use of bamboo reinforcement in lieu of steel rods appears to be by the Szechuan–Hankow Railway, China, in the year 1919[5]. A saving of 50% in cost is said to have been effected by the use of bamboo reinforcement in concrete piles. Bamboo reinforcement was also reported to have been used in a floor in a hospital in Canton, China, and in road pavements[36]. In 1942, several bamboo-reinforced concrete warehouses and gun mounts were constructed by the U.S. armed forces in the Aleutian islands. These structures were designed for a 3-year lifespan. While the actual life period is not known, bamboo is said to have solved a temporary, but critical reinforcement problem. The Japanese Navy is also reported to have used bamboo-reinforced concrete on a large scale, in its military structures during the Second World War. However, no data on these constructions are available.

During 1943–44, three experimental structures, a planer shop, a press box and a five-roomed residence, were built at the Clemson Agricultural College, South Carolina, U.S.A[6]. Different structural elements, such as simple and continuous beams of rectangular and T-shapes, precast slabs, flat slabs, load-bearing walls, columns, and spread footings, all utilizing bamboo reinforcement, were employed. Different treatments for the bamboo were also used. In the planer shop building and the five-roomed residence, distress was observed in the roof beams very early, necessitating local strengthening. But the other portions of the structures were observed to be performing very satisfactorily, some $3\frac{1}{2}$ to $4\frac{1}{2}$ years later. On the other hand, the press box structure, which was three-storeyed, was in good shape some 15 years later, when it was demolished to enlarge the stadium. Purushotham[37] reported the construction of several experimental structures at the Forest Research Institute, Dehra Dun, India. These include a water tank, a road, an arched bus shelter, fence posts, and slabs. In addition to bamboo reinforcement, wooden chips were used in the construction of these structures, with a view to prevent the swelling-shrinking of bamboo. Narayana and Rehman[11], and Cox and Geymayer[15] have also reported constructions for field investigations, but no followup data have been reported.

At the Central Building Research Institute (CBRI), Roorkee, India, a roof slab, 3.6 m × 7.2 m in size, was constructed in 1962, using seasoned and treated bamboo reinforcement with integral shear connectors[38] (Fig. 4.16). A number of cracks, upto 1 mm in width, developed within 3 months after construction. In 1974, these cracks were sealed with 1:1:6 cement:lime:sand mortar. Inspection of the reinforcement at this time revealed no deterioration, and the slab passed the load test recommended by the relevant Indian standards. Today, after nearly 18 years of service, the structure continues to perform well. Between 1974 and 1976, several bamboo-reinforced roof slabs (with spans up to 3.2 m) have also been constructed at CBRI for different store buildings and a dispensary[38,39]. But for a few hairline cracks along the main reinforcement, no other distress has been observed.

Masani *et al.*[19] reported the construction of over 26 experimental and prototype structures using bamboo reinforcement, at the Forest Research Institute, Dehra Dun, India. These structures, built between 1967 and 1971, included different structural units such as roof slabs up to 4.2 m in span; rectangular and T-beams upto 6.2 m in span over 140 lintels and sun shades over door and window openings, cantilevered porticos, picnic sheds, etc. with spans up to 4 m., fence posts, and electric transmission posts. Savings of the order of 30 to 61% are reported, over mild steel reinforced members. Masani *et al.* also reported a destructive load test and long-term load test on bamboo reinforced T-beams of 8.5 m span.

In Manila, Philippines, the beams and slabs of an experimental room (4.2 m × 3.0 m) built with seasoned and treated bamboo splints were observed to be in good condition at the end of seven years[4]. In the Philippines, bamboo reinforcement is recommended for small building structures, and for secondary structural elements such as septic tank covers, kitchen shelves, and pipes. A case of failure has been reported[4] from Indonesia, where the roof of an experimental house constructed with lime-pozzolona concrete and reinforced with bitumen coated bamboo reinforcement suddenly developed serious cracks after six years. However, the reasons for this failure are not known. Recently a bamboo-reinforced concrete pavement has been built in Thailand and opened to traffic in December 1974. This pavement is said to have cost only two-thirds the cost of an equivalent steel-reinforced pavement. It is reported to be giving good service[40].

Bamboo reinforcement can be advantageously used as a substitute for steel bar reinforcement, especially in the developing countries, where steel is scarce and costly. Even where steel is not scarce, bamboo can be gainfully used to reinforce secondary structures, non-load-bearing structures, and structures subjected to low stress levels. Bamboo reinforcement has considerable potential for constructions required for military operations. It has an even a greater role to play in disaster relief structures and in housing for the economically weaker sections[41,42] of the community.

4.3.2 *Bamboo reinforced soil-cement*

Soil-cement is a low-cost material, which is extensively used in pavements, as masonry blocks in low-cost houses, and in other less demanding applications.

Table 4.5 Strength and modulus of soil*-cement (Ref. 14)

S. No.	*Cement content*	*Curing time*	*Compressive strength (wet)*	*Flexural strength (wet)*	*Modulus of elasticity*	*Volumetric shrinkage*
—	%	*Days*	*N/mm²*	*N/mm²*	*kN/mm²*	%
1.	7	14	2.84	0.70	7.59	—
2.	10	14	3.93	0.87	8.97	—
3.	15	14	4.44	1.12	11.73	—
4.	7	28	2.93	0.77	10.35	5.54⁺
5.	10	28	4.83	1.10	11.73	5.07⁺
6.	15	28	6.88	1.32	15.18	2.43⁺

*Alluvial soil with plasticity index = 8.6, and sand content = 40%
⁺After complete curing (initial water content = 10%)

Any soil can be used for soil-cement, provided its plasticity index is between 6.5 and 11.0, and the sand content is not less than 40%. Low cement consumption (typically 5% to 15%), and the use of locally available soil instead of the more expensive aggregates, result in significant economies in the use of soil-cement. Typical strength and elastic modulus values for soil-cement are given in Table 4.5.

A cheap reinforcing material such as bamboo can be economically used to extend the usefulness of soil-cement in structural applications. Also the use of bamboo reinforcement has fewer limitations when used with soil-cement, than with cement concrete. The durability and life of treated bamboo is of the same order as for soil-cement. The thermal coefficients of expansion of bamboo and soil-cement are also not significantly different[43]. Further, the modular ratio for bamboo reinforcement with respect to soil-cement is larger, leading to better utilization of the reinforcement. Also, the shrinkage of soil-cement due to curing is of the order of a few per cent (Table 4.5). Thus, when presoaked and treated bamboo reinforcement is used, the chances of cracking are significantly reduced and the bond between bamboo and soil-cement is not affected.

Table 4.6 Bond strength between soil-cement, and presoaked and treated bamboo (Ref. 14).

S. No.	*Cement content*	*Reinforcement*	*Bond strength*	
			*A**	*B***
—	%	%	*N/mm²*	*N/mm²*
1.	7	1.25	0.62	0.48
2.	10	1.25	0.70	0.70
3.	15	1.25	1.17	0.89

Note: *A—After curing for one week.
**B—After curing for one week, and after subjecting to 12 wetting and drying cycles.
Soil characteristics are as given in Table 4.5.

Coating bamboo with 40% rosin-alcohol mixture followed by a coat of white lead paint has proved to be an effective waterproofing treatment[14,43]. Mehra *et al.*[14,44] observed that the bond strength was affected only slightly due to alternate wetting and drying cycles (Table 4.6). Nainan and Kalam[43] reported similar bond strength values (0.6 to 0.92 N/mm^2) for treated bamboo in soil-cement with 10% cement content. If needed, significant improvements in bond characteristics can be achieved by the use of anchorages, or by suitable surface preparation. Nainan *et al.* obtained improved bond strengths (0.9 N/mm^2 to 1.17 N/mm^2) using techniques such as winding coir rope around the bamboo reinforcement, driving nails into the reinforcement, and sprinkling sand on treated surfaces. The intermediate and end anchorage devices proposed by Datye *et al.*[27] (Fig. 4.11) were highly effective and were able to transfer loads equivalent to half the tensile strength of bamboo, in tests on low-modulus concretes. Further improvements appear to be possible, and could result in the successful use of bamboo reinforcement in soil-cement.

The behaviour of bamboo reinforced soil-cement beams has been investigated by Mehra *et al.*[14,44], Chadda[29,45], Nainan and Kalam[43], and more recently by Pon *et al.*[46] In order to prevent adverse cracking, it has been found necessary[14,44,45] to restrict the tensile reinforcement to about 1.5%, and to adopt proper covers and reinforcement spacings. Typical results of tests on bamboo-reinforced soil-cement beams are presented in Table 4.7. From these and other reported results, it appears that the cracking loads are not significantly affected by either the reinforcement percentage, or the cement content in the soil-cement. On the other hand, increasing cement contents resulted in increasing failure loads, perhaps due to improvement in bond. In almost all the tests reported on beams[14,43–46], the failure was either in bond or in shear. Even in the latter cases, poor end anchorage and poor local bond could have caused the apparent shear failures. Adoption of measures to improve bond characteristics of the bamboo reinforcement has resulted in some improvement. Thus, with the use of shear connectors (of an unspecified type), flexural cracks developed at close intervals in some of the tests of

Table 4.7 Strength of bamboo reinforced soil-cement beams (Ref. 14)

S. No.	*Cement content*	*Reinforcement at*		*Moment at*	
		bottom	*top*	*Cracking*	*Failure*
—	%	%	%	*N.m*	*N.m*
1.	7	1.5	0.0	270.2	391.2
2.	10	1.2	0.0	205.8	425.1
3.	15	1.4	0.0	270.2	573.2
4.	7	1.0	1.0	370.8	492.9
5.	10	1.0	1.0	304.1	648.9
6.	15	1.0	1.0	290.6	831.0

Note: Samples cured for 4 weeks.

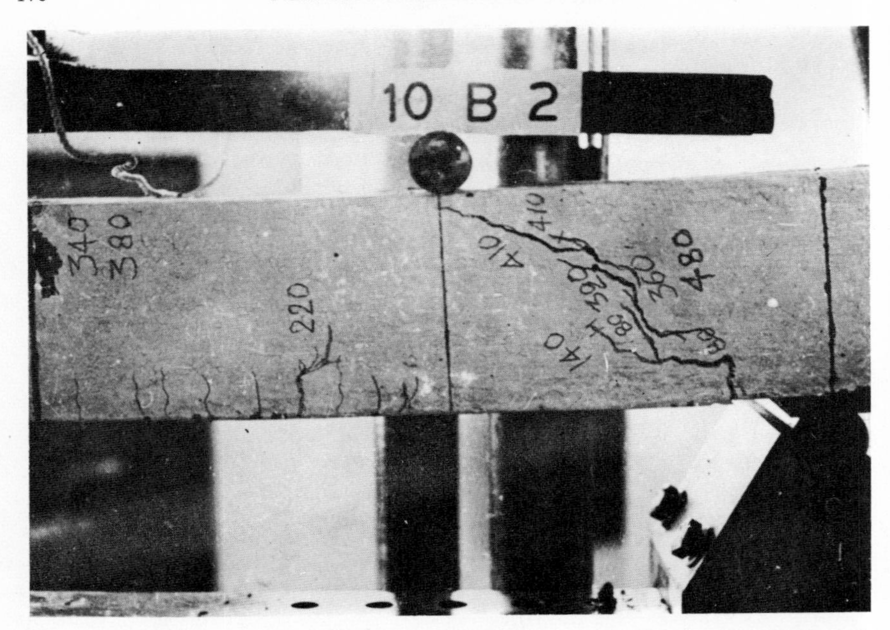

Figure 4.20 Crack pattern in a bamboo reinforced soil-cement beam from Ref. 46. Note the development of flexural cracks at close intervals. (Courtesy of Mr. K.S. Pon, A.C. College of Engineering and Technology, Karaikudi, India)

Pon *et al.*[46] (Fig. 4.20). Using intermediate anchorages, Datye *et al.*[27] obtained improved performance in beams of low modulus concrete (cement content = 40 to 80 kg/m^3 of concrete; and bentonite content = 10% to 30% of cement by weight).

Considering the limitations imposed by bond, and the inelastic nature of soil-cement, Mehra *et al.*[14,44] have proposed the equation for the ultimate flexural capacity, M_u, as

$$M_u = A_b f'_b (d - 0.5n) \tag{4.4}$$

where A_b = area of bamboo reinforcement
f'_b = stress in bamboo at bond failure
d = effective depth of reinforcement
n = neutral axis depth = $A_b f'_b / f_c b$
f_c = average compressive stress
and b = width of the member

This method has been found to be accurate in predicting the failure loads[14,43,44]. A factor of safety of 2.2 to 3.0 is generally recommended for the service moments.

Unlike bamboo-reinforced concrete beams, bamboo reinforced soil-cement beams possess considerable capacity for plastic deformation at failure loads

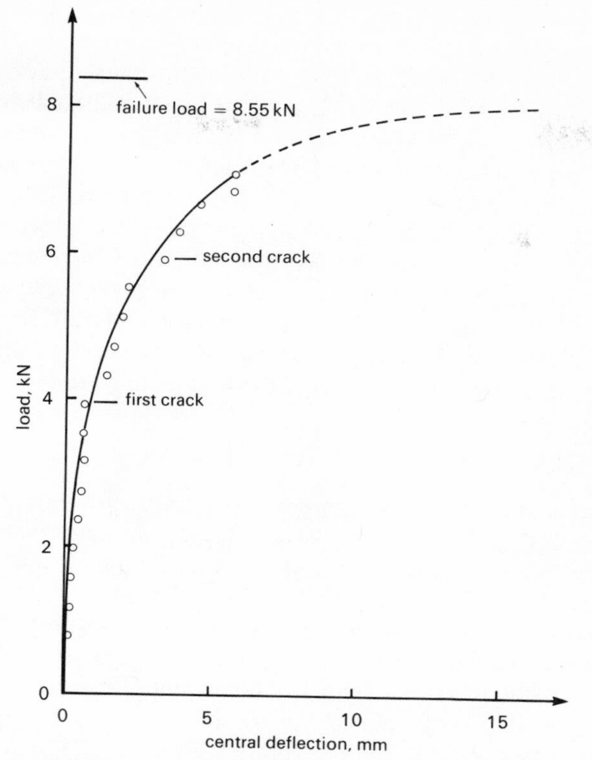

Figure 4.21 Flexural behaviour of bamboo reinforced soil-cement beams (after Mehra *et al.*[14])

(Fig. 4.21), possibly due to the redistribution of the stresses in soil-cement in the compression zone[46]. Analyses[14,43] have shown that the use of bamboo reinforcement results in cost reduction of the order of 30% to 40% over unreinforced soil-cement flexural units. The possible use of bamboo reinforced soil-cement lintels in low-cost houses was first investigated by Chadda[45], and more recently by Nainan and Kalam[43]. The limited investigations of Nainan and Kalam[43] have shown that bamboo is not effective as a compression reinforcement in soil-cement columns. On the other hand, bamboo reinforcement was found to impart considerable rigidity to soil-cement pavements, and continuous footings. This confirms the earlier observations of Mehra *et al.*[14,44] on pavements with bamboo reinforced soil-cement underlays. Bamboo reinforced soil-cement has been used as a bonded underlay in a 270 m long section of a busy road in Rohtak, India, and the performance has been reported to be satisfactory[44,47]. The savings over conventional, plain cement concrete pavement was of the order of 30%[44]. Bamboo reinforced soil-cement has also been suggested[48], for use as flexible raft under embankments on soft soils. Chadda[29] observed that shrinkage cracking in soil-cement walls could be prevented by the use of bamboo reinforcement, at the centre and top of the wall. Bamboo-reinforced soil-cement walls, up to 6 m in length, 2.1 m in height, and 0.3 m in thickness could be built, without shrinkage cracks.

In many developing countries, there is a great potential for the use of bamboo-reinforced soil-cement, in foundations, walls, lintels, and other elements of low-cost houses. In these countries, the use of mud walls is a well-known practice, and only few additional skills may be needed for using the more durable and cost-effective, bamboo reinforced soil-cement. Datye *et al.*[27] suggest the use of bamboo-reinforced low-modulus concrete in foundation rafts, beams for cellular foundations, pavements, aprons, buried conduits and underground pipes. In spite of the considerable interest shown and the potential that exists, applications of bamboo-reinforced, soil-cement and low-modulus concrete are yet to be made, even on a small scale.

4.3.3 *Bamboocrete*

Ferrocement is one of the novel composite materials with a great potential for socially relevant applications in the developing nations, such as low-cost housing, hygienic storage of grain and water, and exploitation of alternative energy sources[49]. The reinforcement for ferrocement, consisting of skeletal steel and wire mesh, costs a significant portion of the total cost of ferrocement. Bamboo reinforcement can be used to replace the steel reinforcement in ferrocement, to obtain a similar construction material at considerably reduced cost. Kalita *et al.*[50] have named the resultant composite material 'Bamboocrete'. Bamboocrete is thus a composite material comprising thinly-formed cement mortar matrix, which is highly reinforced with meshes of suitable, small-sized, bamboo splints.

4.3.3.1 *Structural behaviour of bamboocrete.* Ali and Pama[22] investigated the structural response of bamboocrete in compression, tension, bending, and shear. The initial modulus of elasticity and behaviour of bamboocrete are very similar to those of plain cement mortar. Bamboo reinforcement, in view of its lower elastic modulus and possible buckling of fibres, does not improve the compressive strength of the composite. Indeed, with increasing reinforcement content, there is a falling tendency in the compressive strength (see Fig. 4.22*a*).

In tension and in bending, bamboocrete exhibits a linear response till cracking, which is followed by another linear response, but with significantly reduced stiffness (Fig. 4.23). Ali and Pama observed that the theory of composite materials was applicable, and derived the following equations on the basis of the law of mixtures.

In tension:

(a) Before cracking

$$E_1 = V_m E_m + V_b E_b \tag{4.5}$$

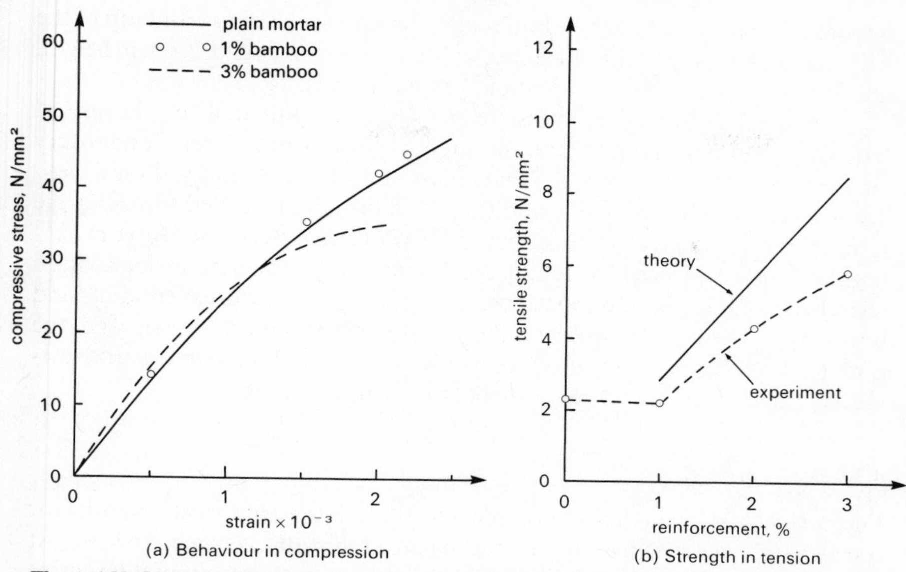

Figure 4.22 Strength of bamboocrete in compression and tension (adapted from Ali and Pama[22])

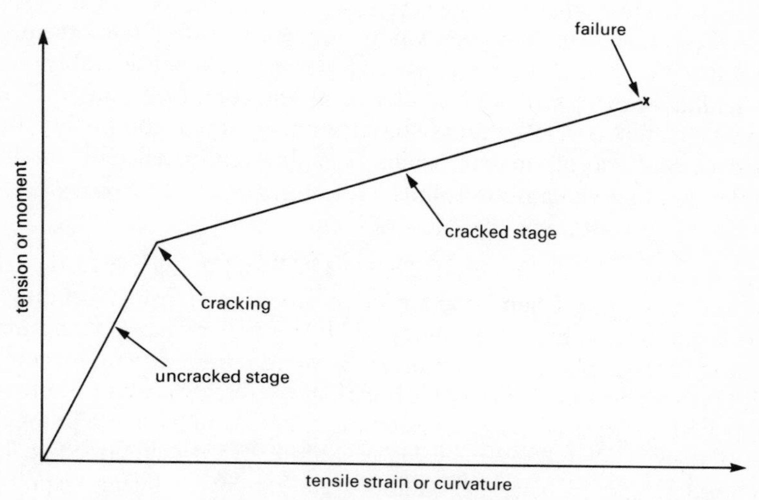

Figure 4.23 Force–deformation response of bamboocrete in tension and flexure

(b) After cracking

$$E_{cr} = \alpha_1 V_m E_m + V_b E_b \tag{4.6}$$

(c) At failure

$$f_{tu} = V_b f_{bf} \tag{4.7}$$

In bending:

(a) Before cracking

$$E_1 = V_m E_m + V_b E_b \tag{4.8}$$

(b) After cracking

$$f_c = \frac{3(1+\beta')}{\beta'} \frac{M}{bh^2} \tag{4.9a}$$

$$f_t = 3(1+\beta') \frac{M}{bh^2} \tag{4.9b}$$

(c) At failure

$$M_u = 0.85 f'_c V_m \frac{h_0^2}{2} + \frac{1}{3} f_{bf} V_b (h - h_0)^2 \tag{4.10}$$

where E_1, E_{cr} = elastic moduli of the composite, before and after cracking

E_m, E_b = elastic moduli for the matrix and the bamboo

V_m, V_b = volume fractions of the matrix and the bamboo reinforcement.

α_1 = factor for reduced contribution of matrix (equal to zero, for advanced stages)

f'_c, f_{bf} = strengths of matrix in compression, and of bamboo in tension

f_c, f_t = stress in compression and in tension

f_{tu} = failure strength of composite

$\beta' = \sqrt{E_1/E_{cr}}$

M, M_u = moment at any stage, and ultimate moment respectively.

b, h = width and thickness of bamboocrete

h_0 = neutral axis depth at ultimate condition.

$$= h \times \left(\frac{0.5 V_b f_{bf}}{0.85 V_m f'_c + 0.5 V_b f_{bf}} \right)$$

Reasonably good agreement was reported[22] between theory and the observed behaviour (Fig. 4.22*b*).

Kalita *et al.*[28,50,51] performed full-scale load tests on bamboocrete wall panels and cylindrical shell roof elements. These elements exhibited linear responses till the cracking of the matrix; this was followed by another near-linear response, albeit with reduced stiffness. From functional considerations, the cracking load may be taken as the usable load carrying capacity of the elements.

4.3.3.2 *Applications.* It is an established practice in countries like India, Indonesia, Peru, Chile, and Ecuador[4] to use bamboo lath as a base for mud, cow-dung, or lime mortar plaster, for walls in low-cost housing. With the

Figure 4.24 Two-roomed bamboocrete house at the Regional Research Laboratory, Jorhat, India. Note the cylindrical bamboocrete roof elements. The walls are also of bamboocrete. (Courtesy of Dr. G. Thyagarajan, Director)

recent development of simple machines in Indonesia to split bamboos into uniform strips and weave them[52], bamboo-reinforced cement mortar (or bamboocrete) is bound to become popular in durable and low-cost constructions, in the developing countries. A significant step in this direction is the development by Kalita *et al.*[28,50,51] of a curved bamboocrete roofing element which could withstand loads up to 2.5 kN/m^2. Bamboocrete wall panels withstood transverse loads upto 5 kN/m^2, before failing. Figure 4.24 shows a full scale bamboocrete house (plinth area = 33 m^2; total cost = Rupees 7100 or US $650), constructed in 1976 by the Regional Research Laboratory, Jorhat, India. This house and a cycle-shed, roofed with bamboocrete, are giving good service.

Bamboo reinforcement replacing only the skeletal steel in ferrocement was successfully employed by Smith[53], and Biggs[54], in building low-cost grain storage silos in Thailand and Bangladesh, respectively.

It appears that bamboocrete can be used for almost all the applications where ferrocement has been used hitherto. Indeed, the use of bamboocrete in lieu of ferrocement is bound to significantly lower the cost. Applications, which are either current or can be foreseen, include wall and roof elements for low-cost housing; utility buildings; service units; silos for granular solids such as grain, fertilizer, and sugar and miscellaneous applications such as dustbins, water troughs for cattle, man-hole covers, inspection covers and cable ducts.

4.3.4 *Bamboo fibre-cement composites*

Asbestos cement sheets and boards are very popular for industrial roofing and for semi-permanent roofing in low-cost houses. In most of the developing countries in Asia, asbestos fibres are scarce, and have to be imported at considerable cost. There is also an increasing, world-wide awareness of the health hazards associated with the use of asbestos fibres. Extensive research is therefore under way in many countries, for investigating the possible use of locally available vegetable fibres instead of asbestos. Among several other

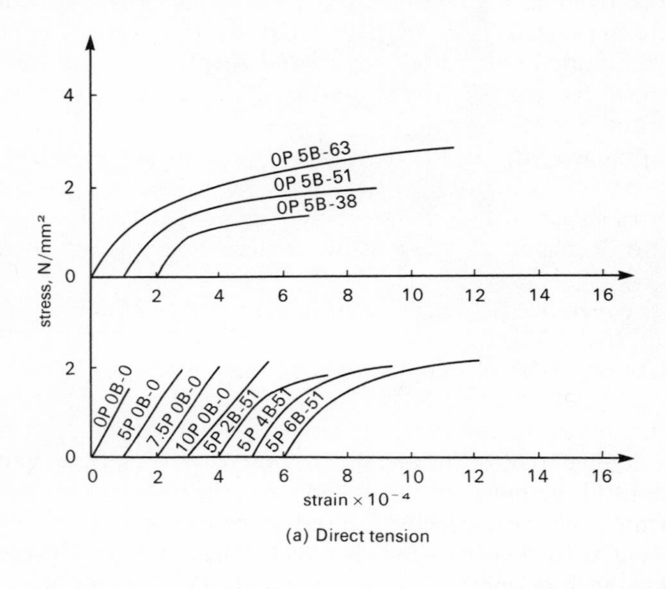

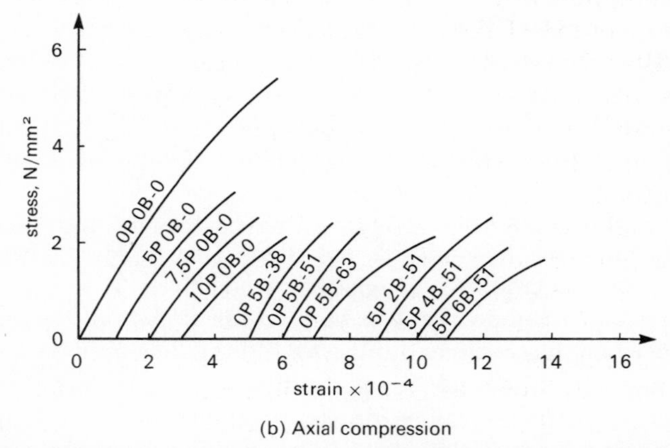

Figure 4.25 Stress–strain behaviour of bamboo fibre–cement composites (after Pakotiprapha *et al.*[55]). *XPYB-Z* denotes a composite with *X* volume percent of pulp, and *Y* volume percent of *Z* mm.-long bamboo fibre)

natural fibres, bamboo fibre also appears to have a good potential as a reinforcement for cement boards and sheets.

Pakotiprapha *et al.*[55–57] investigated in detail the possible use of two types of bamboo fibre to reinforce cement composites. The longer fibres (typically, 3.8 to 6.4 cm in length, and 0.35 mm. in dia.) can be obtained by hammering short sticks of bamboo of specified length. The shorter fibre or pulp (typical length = 0.27 mm; dia = 27.5 microns) is obtained by cooking bamboo chips for 6 hours in 20% sodium hydroxide solution at 170°C, under a steam pressure of about 8.5 atmospheres. The preparation of bamboo fibre-cement composite appears to need certain optimised processing conditions[55,56]. These include a pulp dispersing and mixing time of about 30 minutes, a low water/cement ratio of 0.21, and a casting pressure of 2.5 MN/m^2, maintained for 24 hours.

Pakotiprapha *et al.*[57] have also proposed a method of analysis for bamboo fibre-cement composites, based on law of mixtures. Comparison[56] of the theory with experimental results showed good agreement. However, the influence of increased air voids in the composites needed to be taken into account.

The use of bamboo fibre reinforcement has a general beneficial effect on the behaviour, strength, and ductility of cement paste in direct tension (Fig. 4.25*a*). The degree and type of improvement depends, however, on the type and quantity of the fibre and the increase in air voids. Use of 5 to 10% of short fibre or pulp improved the first cracking strength by about 30%, but the type of failure continued to be brittle and was not altered. On the other hand, the use of the longer fibre resulted in some decrease in the first cracking strength, but subsequently, the composites continued to behave in a ductile fashion till failure occurred at significantly higher loads. The failure loads increased with increase in the fibre length. The difference in the behaviour of the composites has been explained by Pakotiprapha *et al.* in terms of fibre spacing. In the pulp composite, the fibre spacing is less than the critical flaw size. This results in the arrest of microcracks, and the cracking stress is improved. On the other hand, the long fibres act as crack retarders only in the post-cracking stage. Thus, a combination of the two fibres is found to give improved cracking strength as well as ductility (Fig. 4.25*a*)

The flexural strength of cement paste could be more than doubled by the use of bamboo-fibre reinforcement (see Fig. 4.26). As in the case of direct tension, while the short fibre improved the first cracking strength, the longer fibre improved the ductility and ultimate strength. Again, the use of a combination of pulp and long fibre resulted in a superior composite.

The compressive strength of bamboo fibre-cement composites is considerably lower (over 50%) than that of cement paste (Fig. 4.25*b*). The lower elastic modulus and the lower compressive strength of bamboo fibres, and the possibility of buckling of these fibres, are the probable reasons for this behaviour.

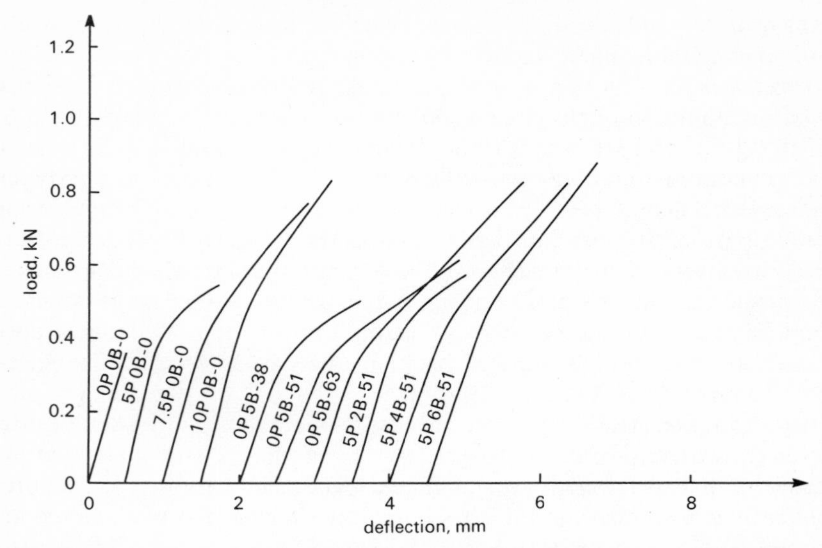

Figure 4.26 Flexural response of bamboo fibre-cement composites (after Pakotiprapha *et al.*[55]). Notation as in Fig. 4.25

Pakotiprapha *et al.* have also performed several tests on the physical characteristics of these composites. In water-absorption and fire tests, the performance of bamboo fibre-cement composites was found to be inferior to that of asbestos-cement sheets. Yet the water absorption of these composites was well within acceptable limits. Also, these composites do not contribute to the inception of fire and are to be classified as non-combustible materials[55]. Impact tests have shown that these composites fail in a ductile manner. Further, permeability tests have indicated that while bamboo fibre-cement boards are relatively more permeable, they still come under the classification of impervious materials.

Thus bamboo fibre-cement composites appear to possess the necessary physical and mechanical characteristics, for use as substitutes for asbestos-cement boards and sheets, for roofing and walling applications. Estimates by Pakotiprapha *et al.* show that savings of the order of 20% to 30% are possible by the use of locally available fibre, in the place of imported asbestos fibre. To fully realize the advantages, long-term and full scale investigations, under field conditions, appear to be necessary. Considering the great demand for lower-cost building materials around the world, there is undoubtedly very good potential for the use of bamboo fibre-cement composites.

4.4 Future outlook

The advantages and the limitations in the use of the different types of bamboo reinforcements are now understood in their proper perspective, thanks to the investigations conducted so far. Aspects needing further investigation are also easily recognized. These include the long-time strength, durability, and structural behaviour of different members built with bamboo-reinforced cement composites. Even though several methods are now available for waterproofing and for prophylactic treatment against deterioration of bamboo, economical and optimal methods are yet to be developed. The different anchorage schemes proposed for improving the bond characteristics of bamboo hold great promise to achieve improved structural performance, and need further study and development. Field applications are also to be made on a larger scale than hitherto, to generate interest and confidence in the use of bamboo as reinforcement. Newer structural forms are also to be developed, using bamboo-cement composites, and the methods of analysis and design are to be further refined. On the other hand, species of bamboo which are more resistant to deterioration and which possess desirable mechanical and other characteristics are to be identified. Such bamboos must be cultivated and harvested in a planned manner. Bamboo reinforcement in its various forms such as whole bamboo, splint, long fibre, or pulp must be standardized and processed in a systematic manner on an industrial scale to make it easily available for field use.

Notwithstanding these future requirements, bamboo reinforced cement composites can be effectively used on the basis of the existing knowledge itself. Bamboo reinforcements can be used with advantage particularly in the developing countries where conventional reinforcements are either scarce or costly. Applications reported so far also indicate the cost advantage to be of the order of 20 to 50%. Considering the spirally increasing demand for housing and other constructions around the world, there is undoubtedly a vast potential for the use of bamboo as reinforcement for cement composites.

Appendix A: Deflection control in BRC beams

Consider a beam subjected to uniformly distributed load, and reinforced with bamboo reinforcement which is effectively treated. Good bond with the matrix is assumed. Under service loads, the maximum deflection,

$$\delta = \frac{5}{48} \frac{M_w l^2}{E_c I} \tag{4A-1}$$

where M_w = maximum moment under service loads
l = span
E_c = modulus of elasticity for concrete
I = moment of inertia

The moment M_w can be written as (see Appendis B)

$$\begin{aligned} M_w &= A_s f_b jd \\ &= (0.9\,\mu f_b) bd^2 \end{aligned} \tag{4A-2}$$

where μ = tensile reinforcement ratio
b = width of beam
d = depth of beam
f_b = stress in bamboo reinforcement under working loads
jd = lever arm $\approx 0.9d$

For the moment of inertia, I, the cracked section moment of inertia I_{cr} can be conservatively used. From Fig. 4B-1a,

$$I_{cr} = \tfrac{1}{3}bn^3 + mA_b(d-n)^2 \tag{4A-3}$$

where $n = kd$
$k = 2/(1+\sqrt{1+2/\mu m})$ (from Appendix B: Eq. 4B-5)

Rewriting

$$I_{cr} = bd^3\left[\frac{k^3}{3} + \mu m(1-k)^2\right] \tag{4A-4}$$

Substituting this value of moment of inertia in Eq. 4A-1, the maximum

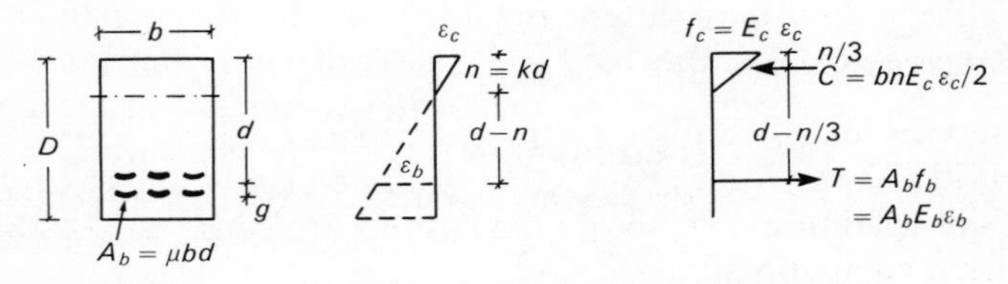

I. Dimensions II. Strain distribution III. Stress distribution

(a) Elastic analysis under service loads

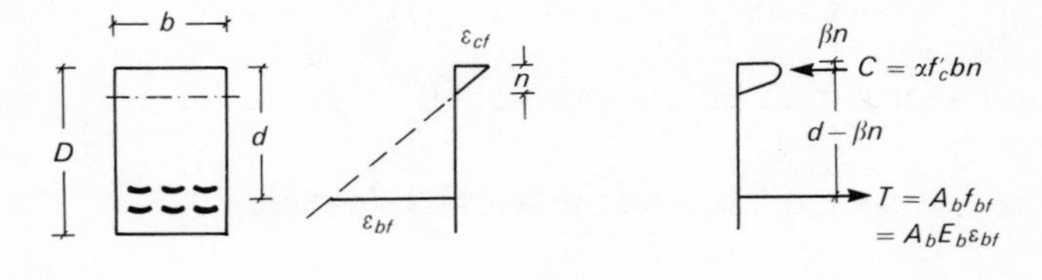

I. Dimensions II. Strain distribution III. Stress distribution

(b) Inelastic analysis at failure load

Figure 4B-1 Analysis of BRC beams

deflection is

$$\delta = \frac{5}{48} \frac{(0.9\mu f_b bd^2)l^2}{E_c bd^3\{(k^3/3) + \mu m(1-k)^2\}}$$

or

$$\frac{\delta}{l} = \frac{l}{d} \times \left[\frac{5}{48} \frac{0.9\mu f_b}{E_c\{(k^3/3) + \mu m(1-k)^2\}} \right] \qquad (4A\text{-}5)$$

For any given limits of acceptable short-term deflection, and given the permissible stress in reinforcement, it is possible from Eq. 4A-5 to calculate the span to effective depth ratio that should be adopted to limit the deflections.

Figure 4.15 gives a typical plot of the span to effective depth (l/d) ratios, if the short-term deflection, δ, were to be limited to a value of $l/360$. It can be seen that the permissible stress in bamboo reinforcement under working loads has a greater influence on these limiting l/d values than the reinforcement percentages. Ratios for other combinations of parameters can similarly be worked out using Eq. 4A-5.

Appendix B: Analysis of BRC beams

(a) *Under service loads*

Under service loads, bamboo-reinforced concrete beams can be analysed using elastic theory, and using the strain compatibility and the force equilibrium conditions. The beam dimensions are given in Fig. 4B-1*a*.

From strain compatibility condition,

$$n = \frac{\varepsilon_c}{\varepsilon_c + \varepsilon_b} \times d \qquad (4B\text{-}1)$$

From force equilibrium condition,

$$C = T \qquad (4B\text{-}2)$$

or

$$\tfrac{1}{2} bnE_c\varepsilon_c = A_b E_b \varepsilon_b \qquad (4B\text{-}3)$$

Substituting for n from Eq. 4B-1 in Eq. 4B-3, and simplifying, we get

$$\varepsilon_b = \frac{\varepsilon_c}{2}[-1 + \sqrt{1 + 2/\mu m}] \qquad (4B\text{-}4)$$

and

$$n = 2d/(1 + \sqrt{1 + 2/\mu m}) \qquad (4B\text{-}5)$$

where μ = reinforcement ratio ($= A_b/bd$)
and m = modular ratio ($= E_b/E_c$)

External moment,

$$M = T \times \text{lever arm}$$
$$= A_b f_b (d - n/3) \qquad (4B\text{-}6)$$

and the stress in bamboo, at the level of the centroid of the reinforcement,

$$f_b = \frac{M}{A_b(d - n/3)} \qquad (4B\text{-}7)$$

However, the maximum stress in the lowermost bamboo reinforcement (see Fig. 4B-1) will be equal to $f_b \times (d - n + g)/(d - n)$.

Example 1

Data: Width of beam, $b = 20$ cm; total depth of beam, $D = 40$ cm; effective depth, $d = 32$ cm; area of bamboo reinforcement, $A_b = 12.8\ \text{cm}^2$; effective span, $l = 4$ m; service moment, $M = 12$ kN/m; elastic modulus of concrete, $E_c = 21.55\ \text{kN/mm}^2$; and elastic modulus of bamboo, $E_b = 18.45\ \text{kN/mm}^2$.

Obtain the stresses and strains under service conditions.
Solution:
For this case,

$$\mu = \frac{A_b}{bd} = \frac{12.8}{20 \times 32} = 0.02$$

$$m = \frac{E_b}{E_c} = \frac{18.45}{21.55} = 0.856$$

The neutral axis depth (from Eq. 4B-5),

$$n = 2d/[1 + \sqrt{1 + 2/(\mu m)}]$$
$$= 2 \times 32/[1 + \sqrt{1 + 2/(0.02 \times 0.856)}]$$
$$= 5.4\ \text{cm}.$$

The stress at the centroidal level of the bamboo reinforcement (from Eq. 4B-7),

$$f_b = \frac{M}{A_b(d - n/3)} = \frac{12 \times 10^5}{12.8 \times (32 - 5.4/3)}\ \text{N/cm}^2$$
$$= 31.0\ \text{N/mm}^2.$$

Strain in the reinforcement,

$$\varepsilon_b = \frac{f_b}{E_b} = \frac{31.0}{18.45 \times 10^3} = 1.683 \times 10^{-3}\ \text{units}$$

Extreme fibre strain in concrete ε_c is obtained from Eq. 4B-1 as

$$\varepsilon_c = \frac{n}{d-n}\varepsilon_b = \frac{5.4}{(32.0-5.4)} \times 1.683 \times 10^{-3}$$
$$= 0.342 \times 10^{-3} \text{ units}$$

Therefore, the maximum compressive stress in concrete,

$$f_c = E_c\varepsilon_c = (21.55 \times 10^3) \times (0.342 \times 10^{-3})$$
$$= 7.4 \text{ N/mm}^2.$$

Thus under service loads,

f_b = stress in bamboo at the centroid of the reinforcement = 31.0 N/mm^2
f_c = compressive stress in concrete = 7.4 N/mm^2
ε_b = strain in bamboo at the centroid of reinforcement = 1.683×10^{-3} units
ε_c = strain in extreme fibre of concrete = 0.342×10^{-3} units

(b) *At failure loads*

In view of the relatively large strains in concrete at failure, a nonlinear (or inelastic) analysis method is to be adopted. A linear strain distribution across the section is assumed (see Fig. 4B-1*b*). For concrete, the realistic compressive stress–strain characteristics as given by Rüsch and Stoeckl[34] are adopted. (Any other accurate relationship may also be adopted). A linear stress–strain relationship is assumed for the bamboo reinforcement, till either the rupture of the reinforcement, or its continuous slippage occurs.

From strain compatibility condition (see Fig. 4B-1*b*),

$$n = \frac{\varepsilon_{cf}}{\varepsilon_{cf} + \varepsilon_{bf}} \times d \qquad \text{(4B-8)}$$

From force equilibrium condition,

$$C = T \qquad \text{(4B-9)}$$

i.e.

$$\alpha f'_c bn = A_b f_{bf} = A_b E_b \varepsilon_{bf} \qquad \text{(4B-10)}$$

where α = average stress parameter.

From Eqs. 4B-9 and 4B-10, we can get the strain in concrete at failure, ε_{cf}, as

$$\varepsilon_{cf} = \frac{\varepsilon_{bf}^2}{\dfrac{\alpha f'_c}{\mu E_b} - \varepsilon_{bf}} \qquad \text{(4B-11)}$$

Since the average stress parameter, α, itself depends on the value of ε_{cf}, this equation can be solved only by trial and error. The neutral axis depth can then be calculated from Eq. 4B-8.

The failure moment

$$M_f = T \times \text{lever arm}$$
$$= A_b f_{bf} \times (d - \beta n) \qquad \text{(4B-12)}$$

where β = lever arm factor, depending on the value of ε_{cf}.

While the above Eq. 4B-12 gives the rigorous failure moment, for most normal cases, it is possible to assume a reasonable value for the lever arm, and calculate the failure moment approximately. For example, if the lever arm is assumed to be equal to 0.9 to 0.95 times the effective depth, since the neutral axis location is relatively high for bamboo reinforced members,

$$M_f = A_b f_{bf} \times (0.9 \text{ to } 0.95) \times d$$
$$= (0.9 \text{ to } 0.95) \times \mu f_{bf} b d^2 \qquad \text{(4B-13)}$$

Example 2

Calculate the failure moment for the beam in Example 1, given also that $f_{bf} = 60\ \text{N/mm}^2$, and $f'_c = 20\ \text{N/mm}^2$.

Solution:

$$\varepsilon_{bf} = \frac{f_{bf}}{E_b} = \frac{60}{18.45 \times 10^3}$$
$$= 3.252 \times 10^{-3} \text{ units}$$

Substituting the values of ε_{bf}, f'_c etc., in Eq. 4B-11, we get

$$\varepsilon_{cf} = \frac{1.057 \times 10^{-2}}{(54.2\,\alpha - 3.252)}$$

This equation has to be solved by trial and error. Assume $\varepsilon_{cf} = 0.94 \times 10^{-3}$: corresponding to $\alpha = 0.368$ (from Ref. 34)

$$\text{Left side of eq.} = 0.94 \times 10^{-3}$$
$$\neq \text{right side of eq.} = 0.633 \times 10^{-3}$$

Assume $\varepsilon_{cf} = 0.766 \times 10^{-3}$: corresponding to $\alpha = 0.315$ (from Ref. 34)

$$\text{Left side of eq.} = 0.766 \times 10^{-3}$$
$$\approx \text{right side of eq.} = 0.764 \times 10^{-3}$$

Hence the extreme fibre strain in concrete,

$$\varepsilon_{cf} = 0.766 \times 10^{-3}$$

Corresponding value of the lever arm factor, β (from Ref. 34) = 0.352
Neutral axis depth

$$n = \frac{\varepsilon_{cf}}{\varepsilon_{cf} + \varepsilon_{bf}} \times d = \frac{0.766 \times 10^{-3} \times 32}{(0.766 \times 10^{-3} + 3.252 \times 10^{-3})}$$
$$= 6.1\ \text{cm}.$$

Lever arm $= d - \beta n = 32.0 - 0.352 \times 6.1 = 29.85$ cm. Hence the failure moment,

$$\begin{aligned} M_f &= A_b f_{bf} \times (d - \beta n) \\ &= 12.8 \times 60 \times 100 \times 29.85 \text{ N/cm} \\ &= 22.9 \text{ kN/m} \end{aligned}$$

(Note: For the sake of comparison, the approximate failure moment calculated using Eq. 4B-13 is

$$\begin{aligned} M_{f,\text{approx.}} &= (0.9 \text{ to } 0.95) A_b f_{bf} d \\ &= (0.9 \text{ to } 0.95) \times 12.8 \times 60 \times 100 \times 32 \text{ N/cm.} \\ &= 22.1 \text{ to } 23.3 \text{ kN/m.} \end{aligned}$$

It can be seen that the rigorous and the approximate failure moments do not differ by more than 4%. Thus, for most normal cases, the simpler equation, Eq. 4B-13 can be used instead of the rigorous method, for calculating the failure moment.)

Appendix C: Considerations in the use of bamboo reinforced concrete

Besides the factors normally considered in steel reinforced concrete, the following additional aspects need to be considered, for successful applications.

1. The maximum size of the aggregate should be limited to 10 or 12 mm, to avoid difficulties, in placement and consolidation of concrete.
2. Use of low water/cement ratios, higher cement contents, plasticizers, and high early strength cement is beneficial, and should be used where possible.
3. Concrete with a compressive strength of at least 20 N/mm^2 should be used, to achieve reasonable bond strengths.
4. To avoid the swelling-shrinking of bamboo, and the consequent adverse cracking and loss of bond, bamboo must be harvested only when it is mature. Also bamboo reinforcement must preferably be used in split form, after proper seasoning and effective treatment, for water proofing and prevention of decay.
5. Split bamboo reinforcement should be reasonably straight. Its width should not exceed 2 to 2.5 cm.

   While placing, the basal and the distal ends of the reinforcement must be alternated, to obtain a uniform reinforcement area along the length of the member.
6. Split bamboo reinforcement should not be arranged with its concave side downwards, as this could result in the entrapment of air during casting, and could lead to poor bond.
7. Bamboo reinforcement is light in weight and tends to move up or float in concrete, during casting. It should therefore be held down by tying to the forms, or by other suitable means.

8. The clear spacing between bamboo reinforcements and the clear cover should at least be equal to the greater of (i) the width of bamboo splints + 0.75 cm, and (ii) the maximum size of the aggregate + 0.75 cm.
9. Bamboo reinforcement may be spliced by giving an overlap of 16 times the width of the splint or by using mechanical splices of the type described by Datye *et al.*[27]. Such splices should be staggered, and should preferably be not located at sections which are highly stressed.
10. Sufficient camber may be given to bamboo reinforced concrete beams and slabs, to reduce the apparent deflections.

References

1. Chu, H.K. (1914) Bamboo for reinforced concrete. Thesis, Massachusetts Institute of Technology, U.S.A.
2. Anon, (1936) The use of bamboo as reinforcement. *Concr. constr. Engng.*, (London) **31**, 618.
3. Datta, K. (1936) Versuche über die Verwendung von Bambus im Betonbau, (Investigation on the use of bamboo in concrete). *Der Bauingenieur* **17**, 17–27.
4. Anon (1972) *The Use of Bamboo and Reeds in Building Construction.* Publication No. ST/SOA/113, Dept. of Economic and Social Affairs, United Nations, New York, 95.
5. Purushotham, A. (1963) Utilisation of bamboo, *J. Timb, Dry. Preserv. Ass. India* **9**, 2–19.
6. Glenn, H.E. (1950) *Bamboo Reinforcement in Portland Cement Concrete.* Bulletin No. 4, Clemson Agricultural College, Clemson, South Carolina, U.S.A., 171.
7. Austin, R., and Ueda, K. (1972) *Bamboo.* Weather Hill Publishing Co., New York, 686.
8. Cox, F.B., and Geymayer, H.G. (1969) *Expedient Reinforcement for Concrete for Use in Southeast Asia: Report 1—Preliminary Tests of Bamboo.* Technical Report No. C-69-3, U.S. Army Engineers Waterways Experiment Station, Vicksburg, Miss., U.S.A., 135.
9. Fang, H.Y., and Mehta, H.C. (1978) Sulfur—sand treated bamboo rod for reinforcing structural concrete. *New Uses of Sulfur-II*, Advances in Chemistry Series 165, American Chemical Society, 241–54.
10. Mehra, S.R., Uppal, H.L., and Chadda, L.R. (1951) Some preliminary investigations in the use of bamboo for reinforcing concrete. *Indian Concr. J.* **25**, 20–1.
11. Narayana, S.K., and Rehman, P.M.A. (1962) Bamboo-concrete composite construction. *J. Instn. Engrs. India* **42**, 426–40.
12. Youssef, M.A.R. (1976) Bamboo as a substitute for steel reinforcement in structural concrete. *New Horizons in Construction Materials*, Vol. 1, H.Y. Fang (ed.), Envo Publishing Co., Lehigh Valley, U.S.A., 525–54.
13. Fang, H.Y., and Fey, S.M. (1978) Mechanism of bamboo-water-concrete interaction. *Proceedings of the International Conference on Materials of Construction for Developing Countries, Bangkok*, 37–48.
14. Mehra, S.R., Ghosh, R.K., and Chadda, L.R. (1957) *Bamboo-Reinforced Soil-Cement as a Construction Material.* Central Road Research Institute, New Delhi, India, 18.
15. Geymayer, H.G., and Cox, F.B. (1970) Bamboo reinforced concrete, *J. Am. Concr. Inst.* **67**, 841–6.
16. Glenn, H.E. *et al.*, (1956) *Seasoning Preservative and Water-repellent Treatment and Physical Property Studies of Bamboo.* Bulletin No. 8, The Clemson Agricultural College, Clemson, U.S.A., 186.
17. Gupchup, V.N., Jayaram, S., and Sukhadwalla, J.N. (1974) Suitability of bamboo strips as tensile reinforcement in concrete. *Proceedings, 6th Congress of the International Council for Building Research Studies and Documentation (CIB), Budapest, Hungary* Oct. 1974, 464–470.
18. Wu, S.C. (1976) The effect of the cutting rotation of bamboo on its mechanical properties. *New Horizons in Construction Materials.* Vol. 1. H.Y. Fang (ed.), Envo publishing Co., Lehigh Valley, U.S.A., 555–66.
19. Masani, N.J., Dhamani, B.C., and Bachan Singh (1977) *Studies on Bamboo Concrete Composite Construction*, Publ. No. PFRI-164, Controller of Publications, Govt. of India, New Delhi, 24.

20. Singh, M.P.J., and Jain, S.K., *Use of Bamboo as Reinforcement in Concrete Slabs*. Technical Note, Central Building Research Institute, Roorkee, India, 2.
21. Panshin, A.J., and de Zeeuw, C. (1970) *Textbook of Wood Technology*. Vol. 1, 3rd edn., McGraw-Hill Book Co., New York, 705.
22. Ali, Z., and Pama, R.P. (1978) Mechanical properties of bamboo reinforced slabs, *Proc. Int. Conf. Materials of Construction for Developing Countries, Bangkok*, 49–66.
23. Cook, D.J., Pama, R.P., and Singh, R.V. (1978) The behaviour of bamboo-reinforced concrete columns subjected to eccentric loads. *Mag. Concr. Res.* **30**, 145–51.
24. Murthy, D.K., and Deshpande, C.V. (1973) Some new techniques to improve the structural behaviour of bamboo reinforced concrete for low cost rural housing. *Nat. Sem. Materials Science and Technology*, Madras, 211–13.
25. Indian Standards Institution, (1970) *National Building Code of India 1970*. Indian Standards Institution, New Delhi.
26. Datye, K.R. (1976) Structural uses of bamboo. *New Horizons in Construction Materials*. Vol. 1, H.Y. Fang (ed.), Envo Publishing Co., Lehigh Valley, U.S.A., 499–510.
27. Datye, K.R., Nagaraju, S.S., and Pandit, C.M. (1978) Engineering applications of bamboo. *Proc. Int. Conf. Materials of Construction for Developing Countries*, Bangkok, 3–20.
28. Kalita, U.C., Khazanchi, A.C., and Thyagarajan, G. (1978) Bamboo-crete wall panels and roofing elements for low cost housing. *Proc. Int. Conf. Materials of Construction for Developing Countries*, Bangkok, 21–35.
29. Chadda, L.R. (1956) The use of bamboo for reinforcing soil-cement eliminating shrinkage cracking in walls. *Indian Concr. J.* **30**, 200–201.
30. Mansut, M.A., and Aziz, M.A. (1983) Study of bamboo-mesh reinforced cement composites. *Int. J. Cem. Comp. and Lightweight Concrete* **5**, 165–171.
31. Srinivasa Rao, P., and Subrahmanyam, B.V. (1973) Simple equation for crack width limitation in reinforced concrete flexural members. *Indian Concr. J.* **47**, 233–236.
32. Kowalski, T.G. (1974) Bamboo reinforced concrete. *Indian Concr. J.* **48**, 119–121.
33. ACI Committee 318 (1977) Proposed revision to the building code requirements for reinforced concrete. *J. Am. Concr. Inst.* **74**, 1–21.
34. Rüsch, H., and Stoeckl, S. (1967) *Characteristics of Rectangular Compression Zone Under Short-time Loading*. (in German). Bulletin No. 196, Deutscher Ausschuss für Stahlbeton, Berlin, 29–66.
35. Bresler, B. (1960) Design criteria for reinforced concrete columns under axial load and biaxial bending. *J. Amr. Concr. Inst.* **57**, 481–90.
36. Hass, A.M. (1983) *Precast Concrete: Design and Applications* Applied Science Publishers, 117–124.
37. Purushotham, A. (1963) A preliminary note on some experiments using bamboo as reinforcement in cement concrete. *J. Timb. Dry. Preserv. Ass. India* **9**, 3–14.
38. Jaisingh, M.P. (1979) Field construction of concrete slabs with bamboo reinforcement made at CBRI, Roorkee, India. Personal communication.
39. Sharma, J.S., Sofat, G.C., and Lathika, J. (1977) Field trials of new construction techniques: Use for dispensary extension. *Indian Concr. J.* **51**, 1–4.
40. Anon. (1978) Bamboo used to reinforce concrete pavements in Asia. *Transportation Research News* No. 79, Transportation Research Board, Washington, USA, 25–6.
41. Arockiasamy, M., and Vijayaraghavan, K.V. (1976) Low cost housing slum clearance project in Madras city—Case studies, *Proc. IAHS Int. Symp. Housing Problems*, Vol. 1, Atlanta, USA, 180–194.
42. Ramaswamy, G.S. (1977) Research and development on low cost housing, *Proc. Int. Sem. Low Cost Housing*, Madras, Vol. 1, IP3/1–22.
43. Nainan, P.K., and Kalam, A.K.A. (1977) Bamboo—reinforced soil-cement for rural use. *Indian Concr. J.* **51**, 382–389.
44. Mehra, S.R., Ghosh, R.K., and Chadda, L.R. (1965) Consideration as material for construction of bamboo-reinforced soil-cement with special reference to its use in pavements. *Civ. Engrg. Publ. Wks. Rev.* **60**, 1457–1461, 1643–1645, 1766–1768.
45. Chadda, L.R. (1956) Bamboo reinforced soil-cement lintels. *Indian Concr. J.* **30**, 303–304.
46. Pon, K.S., Arunagiri, S., and Ravichandran, S. (1980) Bamboo-reinforced soil-cement composite beams, *Proc. Silver Jubilee Sem. Modern Trend in Rural Engineering*, Muzaffarpur, India.
47. Ghosh, R.K., Phull, Y.R., and Chadda, L.R. (1968) Construction of experimental road length

near Rohtak using bamboo-reinforced soil-cement as underlay and base course. 20 *years Design and Construction of Roads and Bridges*, Vol. 1, Ministry of Transport and Shipping (Road Wing), Govt. of India, New Delhi.
48. Mehra, S.R., and Natarajan, T.K. (1963) The flexible raft as a technique for foundation treatment for highway embankments on soft clays. *Proc. Second Asian Regional Conf. Soil Mechanics and Foundation Engineering*, Vol. 1, Japan, 213–216.
49. Subrahmanyam, B.V., and Karim, E.A. (1979) Ferrocement technology: A critical evaluation. *Int. J. Cement Composites* **1**, 125–140.
50. Kalita, U.C., Khazanchi, A.C., and Thyagarajan, G. (1977) Bamboocrete low-cost houses for the masses. *Indian Concr. J.* **51**, 309–12.
51. Anon, *Studies on the Structural Use of Locally Available Materials like Wood, Bamboo, Ekra, etc. and Agricultural Wastes for Construction of Low Cost Houses*. Technical Note, Regional Research Laboratory, Jorhat, India, 7.
52. Dietz, A.G.H., and Moavenzadeh, F. (1977) Innovative uses of materials for housing in developing areas. *Proc. Int. Sem. Low Cost Housing*, Madras, Vol. 1, 42.
53. Anon. (1973) *Ferrocement: Applications in Developing Countries*. National Academy of Sciences, Washington, 55–59.
54. Bigg, G.W. (1975) Bamboo reinforced ferrocement grain storage silo. *J. Struct. Engrg.*, Roorkee, India **2**, 173–182.
55. Pakotiprapha, B., Pama, R.P., and Lee, S.L. (1976) Development of bamboo pulp boards for low cost housing. *Proc. IAHS Int. Symp. Housing Problems*. Vol. 2, Atlanta, 1096–1115.
56. Pakotiprapha, B., Pama, R.P., and Lee, S.L. (1983) Behaviour of a bamboo fibre—cement paste composite. *J. Ferrocement* **13**, 235–248.
57. Pakotiprapha, B., Pama, R.P., and Lee, S.L. (1983) Analysis of a bamboo fibre—cement paste composite. *J. Ferrocement* **13**, 141–158.

Author Index

Subject Index